"It's hard to imagine a more compelling entry-point to the environmental humanities than this. A view through the muddy lens of mangroves, *Exploring Interstitiality with Mangroves* is a bold and unique response to questions about how best to live amidst the tangle of human and nonhuman worlds."

Nigel Clark, *Lancaster University, UK*

"Working as a conceptual 'filter feeder', navigating through the dense underbelly of Sydney's mangroves, Kate Judith accomplishes a rare feat – brilliantly repositioning important conceptual discussions within complex ecologies to explore them from more-than-human perspectives. This move generates valuable new insights for the environmental humanities."

Matthew Kearnes, *Professor, Environment & Society Group, University of New South Wales, Australia*

"An elegant and compelling examination of this world between sea and land. Judith puts leading theorists from philosophy and environmental humanities in the mud of the mangroves; some become allies, others show their limits in this liminal domain. This work announces a unique and powerful voice in environmental thought."

Simon Lumsden, *Associate Professor of Philosophy, University of New South Wales, Australia*

"Kate Judith's book is a nuanced and poetic read into the intertidal fluid and solid spaces of mangroves and their (and our) thicketed ecocultural associations. These pages respectfully take the reader by the senses, heart, and mind into the complex in-between, always dynamic, always inclusive, always inextricably relational."

Tema Milstein, *Environment & Society Group, University of New South Wales, Australia*

Exploring Interstitiality with Mangroves

Mangroves thrive in intertidal zones, where they gather organisms and objects from land, river, and ocean. They develop into complex ecologies in these dynamic in-between spaces. Mobilising resources drawn from semiotic materialism and the environmental humanities, this book seeks a form of social theory from the mangroves; that is to think interstitiality from the perspective of mangroves themselves, exploring the crafty and tenacious world-making they are engaged in.

Three sections weave together theory, science, and close observation, responding to calls within the environmental humanities for detailed attention to interactions in marginal spaces and those of interpretative tension. It examines interstitiality by considering theories of difference, relationality, and reflexivity in the context of mangrove socioecological materialities, drawing on influential writers such as Michel Serres, Jacques Derrida, Deborah Bird Rose, Donna Haraway, Brian Massumi, and Maurice Merleau-Ponty as theoretical touchstones.

Exploring Interstitiality with Mangroves is a lyrically crafted philosophical analysis that will appeal to scholars, researchers, and students interested in the developing frontiers of more-than-human post-anthropocentric writing, theory, and methodologies. It will be of interest to readers in ecocriticism, environmental humanities, cultural geography, place studies, and nature writing.

Kate Judith is an environmental humanities scholar interested in non-anthropocentric practices of meaning, making, relating, and deciding. She is a senior lecturer at the University of Southern Queensland and has a PhD in environmental humanities from the University of New South Wales, Australia.

Routledge Environmental Humanities
Series editors: Scott Slovic *(University of Idaho, USA),*
Joni Adamson *(Arizona State University, USA) and*
Yuki Masami *(Aoyama Gakuin University, Japan)*

The *Routledge Environmental Humanities* series is an original and inspiring venture recognising that today's world agricultural and water crises, ocean pollution and resource depletion, global warming from greenhouse gases, urban sprawl, overpopulation, food insecurity and environmental justice are all *crises of culture*.

The reality of understanding and finding adaptive solutions to our present and future environmental challenges has shifted the epicenter of environmental studies away from an exclusively scientific and technological framework to one that depends on the human-focused disciplines and ideas of the humanities and allied social sciences.

We thus welcome book proposals from all humanities and social sciences disciplines for an inclusive and interdisciplinary series. We favour manuscripts aimed at an international readership and written in a lively and accessible style. The readership comprises scholars and students from the humanities and social sciences and thoughtful readers concerned about the human dimensions of environmental change.

For more information about this series, please visit: www.routledge.com/Routledge-Environmental-Humanities/book-series/REH

Exploring Interstitiality with Mangroves

Semiotic Materialism and the Environmental Humanities

Kate Judith

Routledge
Taylor & Francis Group
LONDON AND NEW YORK

earthscan
from Routledge

First published 2023
by Routledge
4 Park Square, Milton Park, Abingdon, Oxon OX14 4RN

and by Routledge
605 Third Avenue, New York, NY 10158

Routledge is an imprint of the Taylor & Francis Group, an informa business

British Library Cataloguing-in-Publication Data
A catalogue record for this book is available from the British Library

ISBN: 978-1-032-26091-4 (hbk)
ISBN: 978-1-032-26094-5 (pbk)
ISBN: 978-1-003-28649-3 (ebk)

DOI: 10.4324/9781003286493

Typeset in Sabon
by Newgen Publishing UK

Contents

Maps

Acknowledgements

This work was nourished by Cadigal, Kameygal, and Bediagal Country in the Eora Nation where it grew and found its form. I acknowledge the Traditional Owners of these lands, past, present, and future, thank them for their many generations of custodianship, and recognise their continuing connection to these lands and waters. The doctoral studies which form the research base for this book were made possible by an Australian Research Training Scholarship and a University of New South Wales Top-up scholarship and by the University of Southern Queensland who allowed me flexibility in my work arrangements to pursue this research. These doctoral studies were ably supported by my two generous supervisors, Professor Matthew Kearnes and Associate Professor Simon Lumsden, my wonderfully encouraging reader, Associate Professor Tema Milstein, and the always attentive Professor Karen Lai, all at The University of New South Wales (UNSW). I also acknowledge the important support of Associate Professor Thom van Dooren from the University of Sydney, who provided supervisory assistance through the early stages of my PhD, and Professors Nigel Clark from Lancaster University and Michael Marder from the University of the Basque Country whose feedback as thesis examiners has helped improve this book. This book has also benefitted from the wisdom and encouragement of Professor Vicki Kirby and all the members of the "Mullers" reading group which she convenes at UNSW. Other reading groups which have benefited this project include the Relational Inquiry in the Social Sciences and the Science and Technology Studies reading groups at the UNSW, and the Biopolitics of Science reading group at the University of Sydney. I acknowledge the helpful support and encouragement of the Environment and Society Group at UNSW through their work in progress meetings. The publication costs for this book were supported by the Australian Academy of the Humanities through their Publication Subsidy Scheme. I am very grateful for that financial support which has enabled the Introduction to be published open access. I also thank the team at Routledge for their prompt and caring guidance through the publication process. Many thanks to my partner, Ian Barralet, who drew the three maps, greatly helping to locate the events discussed in the research, and who has supported this work in innumerable other ways, and to my friends and colleagues Dr Helena Bender and Dr Hélène Ahlberger Le Deunff, who have been wonderful thinking partners throughout. Finally, I will remain grateful to the mangrove ecosystems of the Cooks and Georges Rivers who inspired and informed this project.

Introduction

A ragged band of mangroves hugs the shore along the lower reaches of the Cooks River in Sydney. If I walk along the river path looking out to the estuarine edge of the mangroves, I see a dark row of trees and protruding roots extending into the water or mud, depending upon the tide, around one to five metres from the riverbank. Along this river, mangroves have been the cause of some community division. In some places they are spreading into areas of rare saltmarsh vegetation and are being removed.[1] In other sites, local conservation groups are monitoring and tending to them, and sometimes they are mysteriously and illegally poisoned.[2] It is hard to get a sense of their status here in the city suburbs. They are not regarded with historic pride like the grand Moreton Bay figs, or with relief like the old Eucalypts shading the children's playground. Parents are more likely to tell children to stay away from the mangroves because of the putrid smell and filthy plastic bags and bottles collecting in their roots. Yet many care about them and are attracted to their strange otherness, the way they hold a world of life and change within their roots and branches.

Estuarine intertidal zones are highly dynamic places, where ocean, river, land, and sky converge, each bringing different materials, organisms, temporalities, and spatialities into play together. Mangroves are specialists in these liminal, challenging spaces, not only surviving there, but building worlds of extraordinary diversity. This work stays with the mangroves, with their edges that don't settle, with the tensions of their locations, with the many ways of knowing them, and with their marginality and dynamism. Where better to explore a non-anthropocentric approach to thinking interstitiality? One insight from having mangroves as companions is the realisation that the spaces between are not empty or static; taking a different perspective, or a different situation, may find them teeming.

Interstitiality, mangroves, and more-than-human approaches to enquiry

One of the major challenges for the environmental humanities is to explore what it means to understand and create in response to an apparent collapsing

DOI: 10.4324/9781003286493-1

of categories and assumptions. The previously hardened divisions of geology and biology,[3] purity and pollution,[4] natural and artificial,[5] alive and not alive,[6] domesticated and wild,[7] culture and nature,[8] have softened and paradigms that assume human uniqueness and the virtue of human mastery and progress are being challenged.[9] Boundaries that once seemed solid, edges that held firm, are giving way to less well-defined, more ambiguous spaces and uncertain futures. This uncertainty has been described in numerous ways, including as increasing risk that affects our capacity to plan,[10] as physical and psychic pain,[11] and as challenges for thought and politics.[12] Along with the old categories, many theoretical and methodological tools (use value, ecosystems services, sustainability, and resilience, for example) are also being questioned as too tightly bound to assumptions of human uniqueness or system stability.[13] The diverse field of enquiry referred to as the environmental humanities is characterised by an emergent analytic ethics that seeks to overcome the deeply etched grooves of anthropocentric paradigms. Environmental humanities scholars are exploring creative modes of enquiry that do not assume the primacy of human interests and human perspectives, assumptions which are associated with ecological insensitivity leading to environmental destruction.[14] By exploring questions of betweenness within the semiotic materiality of mangroves, this work approaches those vexed dualisms and categories in a way that elides anthropocentrism and aims to contribute theoretically and methodologically towards supporting a research shift away from anthropocentrism and assumptions of human exceptionalism. It does this in two ways: by building conceptual capacity for thinking in-between categories and by adopting a more-than-human perspective.

In-between, liminal, and marginal spaces or states have been identified as key sites where assumptions and hegemonies are both formed and destabilised[15] or where associations, interactions, and connections provide the potential for the emergence of the new.[16] Within interstices, edges and differences are problematic and in-process. Interstices are not empty spaces between established things but are already thick with complex happenings that are likely to be interesting and where comfortable assumptions are not secure. In some contexts, the instability and uncertainty of the interstitial feels strange, monstrous, or impossible – themes this work will be investigating. According to Tim Ingold, while "between" "articulates a divided world", "in-between" is not defined by termini. It has its own status as "interstitial differentiation",[17] which Ingold instantiates with flow midstream and breath that continually inhales and exhales. Ingold suggests we are immersed in the in-between, which is the site of "answering and being answered to",[18] and because of that, the site, ultimately, of responsibility. Likewise, Brian Massumi calls directly for attention to the in-between in its own right, arguing that so long as being between is only considered as the marginal derivative of positive thingness, thinking continues to turn towards foundationalism, and fails to meet the challenges of complexity and inter-affective relationality. "The tendency is to describe the in-between as a blending or parody of the always-already positioned", he explains.[19] Where this occurs, concepts that claim to consider interstitiality, such as marginality or mixture "retain a necessary reference to the pure, the central, and the strait-laced and straightfaced, without which they vaporize into logical indeterminacy".[20]

This research takes up this challenge, selecting theory that has focused on the in-between, and taking it to the mangroves to develop a non-anthropocentric understanding of the uncertain, ambiguous spaces of becoming, relationality, and difference.[21] The work explores concepts from within literature sympathetic to post-anthropocentric material semiotics that have sought to make interstitiality thinkable. I seek out concepts that remain close-up, peering into the details of the performances of interstitially.[22] Through this body of theoretical work, interstices take many guises: sometimes as a generative disturbance or a site of friction or irritation of materiality, other times as a contestation, a time or space of becoming, an entanglement of relations, a bifurcation or fragmentation, or a decision or indecision. Descriptors for the interstitial are equally varied.[23] Some of the writers whose thinking is drawn upon through this study employ dense vocabularies of neologisms to find ways to discuss interstitiality.[24] While there are differences between the approaches considered within this research, the various concepts I will be exploring are all similarly engaged with finding ways of allowing processes of generativity, dissolution, indeterminacy, and tension to be accessible to discussion while respecting their in-betweenness. By bringing these various accounts into conversations with mangrove ecology, I hope to develop a more-than-human exploration of interstitiality with some coherency, complexity, and breadth of significance.

The book progresses from an ontological focus in Part 1, to a focus on relationality in Part 2, to issues of power and politics in Part 3. Each section is organised around one broad concept, these being reflexivity in becoming, relationality, and difference. These provide me with three ways of enquiring into betweenness, three sets of questions and three extensive bodies of theoretical work. My inquiry considers the roles each of these three plays in establishing and maintaining the premises, attitudes, and practices of human exceptionalism. When I consider difference, my underlying intention is to provide insight into difference as it operates between humans and non-humans in those situations where human exceptionalism is problematic and to explore alternative ways of differencing. When I investigate relationality, it is with the recognition that human exceptionalism is maintained through particular ways of understanding and valuing relations, and when I explore reflexivity, it is to understand better how the reflexive moves that fold semiotic materiality into meaningfulness support human exceptionalism and how they might be folded otherwise.

A semiotic material ontology frames my thinking and arguments. In the broad sense in which I understand it,[25] semiotic materialism includes a legacy of writing that attends to its own textual practice, that is self-consciously performative in exemplifying the theory it explores in practice. The text reflects the legacy of writers such as Gilles Deleuze (in Part 1), Donna Haraway (in Part 2), and Jacques Derrida (in Part 3), in regarding itself as performing a material semiotic practice as it explores semiotic material concepts. Important to this project in this regard is its situatedness within a particular semiotic materiality, that of the mangroves of Sydney's southern rivers. It was indeed the mangroves that guided my reading towards thinkers who embrace a semiotic materiality. As I sought thinkers whose work spoke to my experience of tidal rhythms and mangal relations, this was the body of theory that responded.

This book thinks with mangroves, whose preoccupation it is to occupy, complicate, and experiment in-betweenness. My research interlocutors throughout this study are the mangrove communities of the Cooks and Georges Rivers in southern Sydney. These mangroves are specialists in negotiating the space between land and water, salt and fresh, coast and river, change and continuity, regularity and disturbance, as well as negotiating with changing sea levels, tides, floods, and the chemical and material composition of their environments. In mangrove ways, they enact the between, stretching out through interconnections. Humans whose bodies and actions affect and are affected by the mangroves are part of that community, as are the migratory birds who stop there to roost or feed, and the sharks who follow the smaller fish into the safety of mangrove roots. Along the way, this study may follow a thread away from the riverbanks, into human wall building, immunology, or lunar traditions.

Introducing mangroves

Mangroves, a term used to describe both the trees and the forest type, including all the plants and animals as well as the non-living conditions required for their survival, thrive in a band along parts of the intertidal waterline of coasts and estuaries in tropical and subtropical regions.[26] Mangrove (or mangal) plants are taxonomically diverse, comprising some 54–75 species globally.[27] All mangrove tree species have evolutionary adaptations to enable them to survive in anaerobic, saline conditions and unstable waterlogged soil.[28] Most characteristic are the specialised roots. Pneumatophores, for example, are roots that protrude from the mud like small spiky fingers and absorb air during low tides through holes called lenticels. Other specialised mangrove roots act as stilts, forming stabilising networks of scaffolding for the tree trunks. To manage the high salinity, some species exclude salt through filtering processes, while others deal with it by excreting it through their leaves.[29] With these and many other specialised adaptions, mangrove trees are able to grow densely when conditions are suitable, forming a thick dark canopy.

Despite their limited range, squeezed within the tidal zone, mangroves are of enormous importance in many ways. Massive and complex root systems and dense branches accommodate a vast range of organisms, and runoff from the land, tidal wash from the sea and the continual fall of mangrove leaves bring regular supplies of nutrients and other resources into this porous and capacious environment. This combination of regular incoming resources and diverse niches mean mangroves afford many opportunities for animals seeking refuge, food, shelter, mates, or nursery spaces for young.[30] Animals from sea, land, and sky converge here, along with many mangal specialists.

Mangroves were, and in many areas still are, highly valued by Indigenous coastal groups in Australia for their abundance of resources.[31] The Wangal, Cadigal, and Kameygal people of the Cooks River and the Gweagal, Cabrogal, and Bediagal[32] groups along the Georges River cut shields from *Avicennia marina* (commonly known as the grey mangrove) bark, fished with spears, and harvested shellfish, particularly oysters, mussels, and cockles from mud banks and sandbanks in large quantities, and along the Georges, mangrove worms or cah-bro were an abundant and favoured food.[33] Australian mangroves are also culturally and historically important within Indigenous contexts. Knowledge about how to look after the mangal and harvest its foods and other resources remains highly valued, particularly in northern Australia, and the mangrove-edged estuaries of the Cooks and particularly the Georges Rivers were important sites of Indigenous survival following the British invasion.[34]

In recent decades, climate change has drawn increasing attention to the significance of mangroves in coastal protection from storm damage and sea-level rise. Coasts protected by mangroves suffer less damage during storms than those where mangroves have been removed, sometimes far less where the mangroves are thick and healthy.[35] Furthermore, mangrove mud accumulates large amounts of organic matter, which degrades relatively slowly within the anaerobic environment, making mangroves one of the world's most effective environments for storing carbon.[36] These features make mangroves particularly "good to think with" in terms of considering the limits of expertise in approaching climate change, according to Sarah E. Vaughn.[37] As responders to and mediators of climate change, mangroves are moving into another kind of between, one that makes them particularly significant right now.

Cooks and Georges River mangroves

The mangroves of the Cooks and Georges Rivers survive within Sydney's urban environment and have had lengthy and troubled urban histories. Both these rivers run through the southern suburbs of the city and flow out into Botany Bay, to the south of Port Jackson and Sydney harbour.[38] They are ancient, predating the inundation of Botany Bay at the end of the last ice age. Before that inundation, they flowed together to a coast far to the east of its current location.[39] Indigenous occupation along the rivers extends back before that inundation.[40]

The Cooks River[41] is notorious as Australia's most altered and polluted river.[42] Noted by Captain James Cook in 1770 as a bountiful environment providing ample resources for the Indigenous inhabitants and fit to be Australia's first British settlement, the estuarine lower reaches of the low lying, slow-flowing river were fringed with mudflats, saltmarsh, banksia scrub, and extensive mangroves and supported a great diversity of animal species.[43] Cook's opinion was not shared by the officers of the first fleet, landing in 1788 and seeking a suitable site. They discovered a shallow intermittent flow which they judged would be unlikely to provide reliable fresh water for the colony. The city, therefore, grew with its face towards Port Jackson and the Parramatta River. The Cooks and its extensive and productive floodplains, mudflats, mangroves, and saltmarsh became the city's hard-working, "back door".[44] Sydney's rapid development quickly destroyed almost all the Cooks River mangroves. A dam built across the river in 1839–40 blocked tidal flows and killed all the mangroves upstream. Downstream from the dam, the broad areas of mangrove swamps were extensively utilised as a source of potash for soap-making to the point of depletion[45] and were devastated during the extraction of shells from Aboriginal middens for lime production and failed attempts at oyster farming.[46] They then increased their spread in response to the increased supply of nutrients from the city's sewage, some of which was diverted towards Botany Bay. Subsequently, the mangroves became regarded as dangerous sites of criminality and noxious miasma.[47] Eventually, in the 1940s, they were drained and reshaped into Sydney airport extensions and industrial areas. In recent decades, the old industrial areas and working-class suburbs surrounding the river have been gentrifying, river pollution is diminishing, and mangroves have been returning to the river. Community attitudes towards them, though mixed, are shifting towards greater acceptance and even limited encouragement.

The broader and longer Georges River[48] flows into the southern part of Botany Bay. It is approximately 80 kilometres long with a catchment of over 930 square kilometres which includes relatively undeveloped areas of bushland in the upper reaches. Through its middle reaches the Georges meanders through low-lying country which would once have been swampy floodplains but is now sprawling suburbs and significant areas of medium and heavy industry. The river widens into a broad estuary edged in most places by steep-sided sandstone banks which are broken by numerous inlets. It is within these inlets that many of the mangrove communities of the river are to be found. These lower reaches of the river are also surrounded by the southern suburbs of Sydney.

The Georges River mangroves have varied histories, depending upon the land use around each inlet. In several inlets, historical use as rubbish dumps was regarded as actually improving the value of these much denigrated and even feared habitats.[49] These rubbish dumps were later filled and repurposed as sports grounds or open areas. In some other inlets, surrounded by industrial areas or national parks, the mangroves have remained less disturbed. Through the early twentieth century, the Georges was particularly important as a site of Indigenous continuity and resistance.[50] The rugged terrain, with steep-sided escarpments and many inlets, provided areas where "colonial control could not secure",[51] where Indigenous groups could survive and maintain some self-determination. One well-known camping area that continued to be occupied by Indigenous people into at least the mid-twentieth century was at Salt Pan Creek, where mangroves remained near the creek's river entrance. Indigenous groups survived and adapted at this camp alongside Sydney's mangroves.

The mangal of practice: material semiosis as filter-feeding

The search for non-anthropocentric research methodologies

One of my aims with this book is that it might constitute an example of research that disrupts and seeks alternatives to anthropocentrism. Locating mangroves at the centre is an integral part of that aim. Much work in the environmental humanities concerns the development of approaches to thinking, writing, and researching that avoid a solely human perspective. There are numerous approaches to this challenge. One of these involves expanding the limits of agency, communication and thought to include non-humans. A host of recent research is revealing the ways that, for example, plants and fungi communicate with each other, and how what we think of as decision, choice or agency cannot justifiably be limited to humans. Another approach is to expand models of participatory research to involve non-humans within the research processes, from design through to dissemination.[52] Beyond environmental humanities, methodologies contributing towards this challenge include multispecies ethology and ethnography, which have shifted their research fields towards considering humans as entangled within multispecies relations.[53] Within my work, I am asking how concepts that address interstitiality adapt and change as they are filtered through mangroves. What happens to these concepts when they wash through mangroves, how do mangroves change them? How can the mangroves with their muddiness and sponginess and shifting rhythms speak back to theory?

An alternative word that describes all things mangrove is "mangal", which can be used as an adjective or a noun, as can the word "mangrove". Serendipitously, it is the homophone of "mangle", the term for the device used prior to the manufacture of washing machines to wring water out of washing and also the verb for doing that squeezing. Andrew Pickering chose the mangle to help him describe the process through which "material and human agencies are mutually and emergently productive of one another" within scientific practice.[54] He called this process "the mangle of practice", to emphasise the performative and posthuman qualities of the process through which the products of science and human intellectual activity emerge unpredictably from practices that involve human intentionality always mangled together with the agencies of the material world. Pickering's mangle of practice overtly assumes a separation of human intentionality from non-human materiality, which become mangled together within practices that occur within specific contexts: the laboratory, the engineering workshop, or academic enquiry. While useful for thinking about the construction of knowledge within these contexts, it arguably does not sufficiently acknowledge the extent to which human intentionality and material agencies are already thoroughly mangled before they get into these spaces and therefore seems to be applicable in contexts that assume an anthropocentric position. Perhaps some of the trouble lies with Pickering's choice of mangle. He focussed on a form of mangle that has the effect of emphasising human practices, the kind that is worked by a human and defined by its functionality towards human purposes. He applies a mangle-machine-metaphor to practices, assuming they are principally performed within the semiotic parameters of a human-centred context of meaningfulness, and never escapes the loop of anthropocentrism. Mangling does not escape the forceful imposition by humans of human-orientations upon what is brought violently together. In this work, I ask: "what might it mean to think with a mangrove understanding of the mangal?"

My hope for this project is that it can perform a "mangal of practice" that does not require an exclusively human-centred epistemological universe from which questions step off, or into which the response is presented for judgement. Here the questions asked, the work done through the practices of walking, flowing, listening, differentiating, processing salt, reading, photosynthesising, thinking, flocculating, writing, infecting, revising, or filter-feeding, are mangaled within a semiotics that is always ecological and material. The linguistic form of this book is the written English language, the hands that typed the words are mine; mangaling does not deny difference, emergence or semiotic meaningfulness, including meanings that are accessible within particular relations such as those of a human cultural practice. Like Pickering's mangle, mangaling describes practices of becoming, transformation and differentiation; both are about the performances of becoming meaningful. Unlike mangling, mangaling does not suppose these descriptions, practices and meanings necessarily involve an element of human exclusivity and exceptionalism, or that humans must perform a violent squeezing of the human and the material together. The process is more like filter-feeding. A lot of filtering-feeding goes on through a mangrove. Mangal mud is a vast filter itself, and many filter-feeders reside there, gathering their bodies and their durations from tidal current and drifting sediment, and through their filter-feeding, forming the geobiome of the mud. Rather than the violent forcing together of Pickering's technological mangle, this project filter-feeds theory and mangroves together. My filtering apparatus is transformed through the process. This story of material semiotics as filter-feeding will be expanded through the first chapter, "The proposals of tides and the responses of oysters".

The book is organised into three major parts, each part addressing one of the three broad conceptual areas of the work: ontological reflexivity, relationality, and difference. Each is explored through a narrative feature of the Cooks River or Georges River mangroves, and each develops an argument intended to help align theory with the needs of non-anthropocentric, more-than-human futures.

Part 1, "Strange reflexivities: folding in communicative tidal materialities", considers tides and filter-feeding to explore reflexivity as it is involved in becoming. Its questions are ontological, and it dwells within tidal rhythms and the filter-feeding processes of the mangroves. Oyster bodies have responded to tidal rhythms over evolutionary time and in everyday interactions to work with the flow of the water current and the suspended particles. Selecting, excluding, digesting, and excreting, oysters nourish and build fleshy bodies and hard-shell reefs, filter water, and produce the geobiome of the rich mangrove mud. The three chapters develop a discussion, worked through mangrove materialities, comparing ontologies that propose a semiotic materiality that is meaningful because it comprises a nested hierarchy of layers of increasing semiotic abstraction and those that propose a folded or involuted semiotic materiality. It asks whether the thick gap of semiotic material meaning-making is located within a reflexive shift that establishes levels of abstraction, or whether reflexive processes are dispersed, unstable, and indecipherable as layers. I argue that hierarchically layered semiotic material ontologies have tended to perpetuate the legacy of anthropocentrism and that folding and involution offer ways of thinking the reflexivity involved in meaning-making more suited to the needs of post-anthropocentric theory.

Part 2, "Monstrous relations: exploring a hermeneutic account of relationality", investigates a distinct way of thinking relationality. These three chapters follow the unfolding relations extending between a wallaby, an *Aedes vigilax* mosquito, Ross River Virus, and my body. Our relations have an epic quality – they are rich with life and death risks, sickness, desire, hunger, exploitation, and the nourishment of new generations. Our bodily exchanges are, as Beisel, Kelly, and Tousinant proclaim, "momentous"[55] – and while there are many attunements between us, these often seem to occur agonistically. Most of this discussion focuses upon our encounters one summer around the mangrove edged bays of the Georges River. My researcher's body is reflexively enfolded, called into questions of troubled co-becomings. My visits to the river become layered with an awareness of my body as food for future generations, as a complex experimental field of immunity and accommodation, and as a part of an ancient field of lively experimentation that leaps between, extends across, and shapes all our different bodies.[56] The key proposal of this section is that in a semiotic material ontology, relations are narratives, and therefore narrative strategies are powerful when employed selectively within relationships. I argue that conceptualising relations as narratives within a semiotic materiality is both theoretically coherent and practically useful for guiding humans in the challenges of living better among others. Such a conceptualisation enables a diverse range of hermeneutic-relational methodologies to be considered for their efficacy in critiquing, supporting, or disrupting existing relations and working new or revised relations, in response to these challenges. In Australia, as in many other places where First Nations people maintain deep connections to land through practices that respect the agency and communicative capacity of other-than-humans, our paths towards more ecologically attuned practices should follow First Nations guidance. Many of the narrative strategies considered here, therefore, have their origins in First Nations practices.

Part 3, "Impossible differences: a muddy journey across more-than-human walls and hospitalities", focuses upon difference in relation to

Map 0.1 Map of the New South Wales coast from Port Jackson to Port Hacking, showing the relative positions of the Parramatta River, Cooks River, Georges River, and Hacking River.

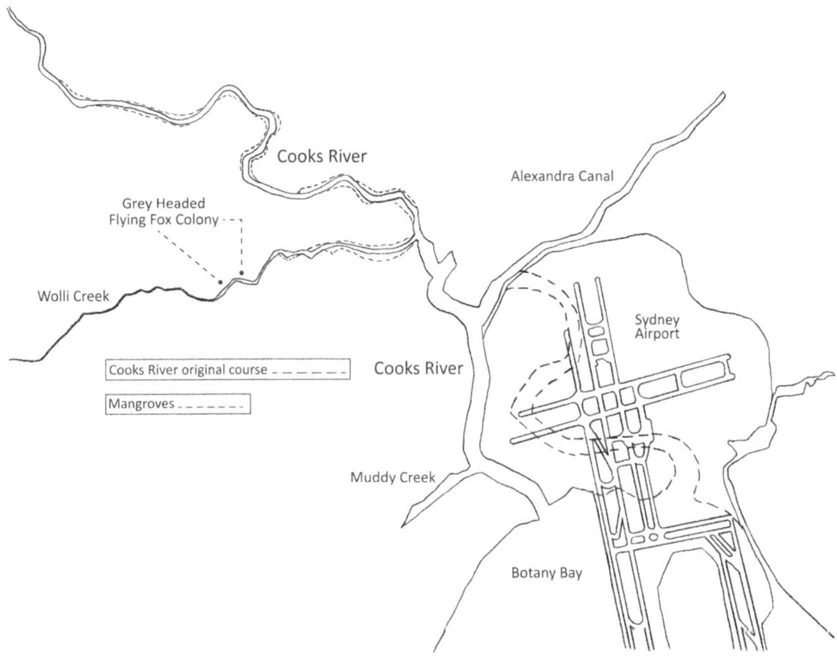

Map 0.2 Map of the Cooks River showing the areas of mangroves, the location of the flying-fox colony, and the course of the river prior to the construction of Sydney Airport.

questions of power, territory, and sovereignty. The history of the relationship between the dam walls and embankments of the Cooks River and the river's mangroves provides the mangal materiality for this investigation. The key proposals here are that it is important to understand the attractiveness of securing a powerful position, and that a thick, folded, relational semiotic materiality (as developed through the first two parts of the book, and instantiated by a mangrove) offers an alternative and useful way of reconceptualising security and its artefacts as porous, accommodating, and supportive of a reconceptualisation of vulnerability as a shared resource.

Ecological complexity continues, changes, and diversifies because this world is communicative in multitudes of intersecting ways. Anthropocentrism precludes almost all that communicative capacity from participating in the conversations shaping our common futures. This work contributes towards one of the great tasks of the environmental humanities; how to decentre human perspectives and interests and find ways to continually accommodate the communicative expressions, agencies, and interests of non-humans within decision-making that affects all our entangled futures.

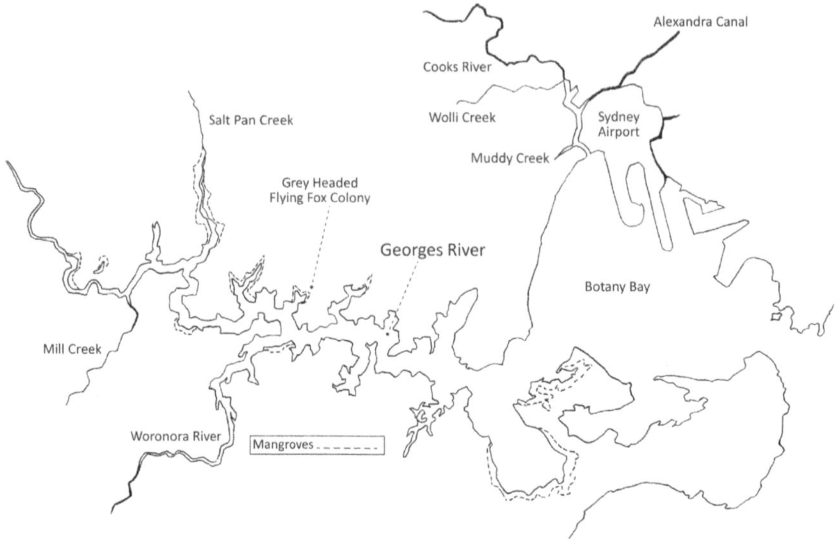

Map 0.3 Map of the lower reaches of the Georges River showing the areas of mangroves, and the location of the flying-fox colony.

Notes

1 Cooks River Alliance, *The Cooks River People's Plan: A Strategic Plan for United Action on the Cooks River and Its Catchment 2018–2021*, https://cooksriver.org.au/wp-content/uploads/2018/08/Cooks-River-Alliance-Strategic-Plan-2018-2021.pdf
2 Jacqueline Yetzotis, "Cooks River Mangroves Poisoned by Vandal," *Saving Our Trees – Marrickville Municipality*, accessed 21 May 2017, https://savingourtrees.wordpress.com/2012/10/16/cooks-river-mangroves-poisoned-by-vandal/
3 Nigel Clark, *Inhuman Nature: Sociable Life on a Dynamic Planet* (Nottingham: Sage Publications Ltd, 2011); Kathryn Yusoff, "Geologic Subjects: Nonhuman Origins, Geomorphic Aesthetics and the Art of Becoming Inhuman," *Cultural Geographies* 22, no. 3 (2014): 383–407, https://dx.doi.org/10.1177/1474474014545301
4 Alexis Shotwell, *Against Purity: Living Ethically in Compromised Times* (Minneapolis and London: University of Minnesota Press, 2016), Kindle.
5 Donna Haraway, "A Cyborg Manifesto: Science, Technology, and Socialist-Feminism in the Late Twentieth Century," in *Simians, Cyborgs and Women: The Reinvention of Nature* (New York: Routledge, 1991); Donna Haraway, *Modest_Witness@Second_Millennium.Femaleman_Meets_Oncomouse: Feminism and Technoscience* (New York: Routledge, 1997): 149–181.
6 Elizabeth A. Povinelli, *Geontologies: A Requium to Late Liberalism* (Durham and London: Duke University Press, 2016), Kindle.

7 Kohn Eduardo, *How Forests Think* (Berkeley and Los Angeles: University of California Press, 2013); Anna Lowenhaupt Tsing, *The Mushroom at the End of the World* (Princeton: Princeton University Press, 2015), Kindle.

8 Val Plumwood, *Feminism and the Mastery of Nature* (London and New York: Routledge, 1993), Kindle.

9 Val Plumwood, *The Eye of the Crocodile* (Canberra: Australian National University E Press, 2012); Deborah Bird Rose, "Multispecies Knots of Ethical Time," *Environmental Philosophy* 9, no. 1 (2012): 127–40.

10 Ulrich Beck, *World at Risk*, trans. Ciaran Cronin (Cambridge: Polity Press, 2009).

11 Lauren Berlant, *Cruel Optimism* (Durham and London: Duke University Press, 2011), Kindle; Judith Butler, *Precarious Life: The Power of Mourning and Violence* (London and New York: Verso, 2004).

12 Donna Haraway, *Staying with the Trouble: Making Kin in the Chthulucene* (Durham and London: Duke University Press, 2016); William Connolly, *Facing the Planetary: Entangled Humanism and the Politics of Swarming* (Durham and London: Duke University Press, 2017), Kindle.

13 Stacey Alaimo, *Exposed: Environmental Politics and Pleasures in Posthuman Times* (Minneapolis and London: University of Minnesota Press, 2016), Kindle.

14 Bawaka Country et al., "Working with and Learning from Country: Decentring Human Authority," *Cultural Geographies* 22, no. 2 (2005), https://dx.doi.org/10.1177/1474474014539248

15 William Connolly, *Politics and Ambiguity: Rhetoric of the Human Sciences* (Madison: University of Wisconsin Press, 1987); Erinn Cunliffe Gilson, *The Ethics of Vulnerability: A Feminist Analysis of Social Life and Practice* (New York: Routledge, 2014), Kindle.

16 Gilles Deleuze and Felix Guattari, *A Thousand Plateaus: Capitalism and Schizophrenia*, trans. Brian Massumi (Minneapolis and London: University of Minnesota Press, 1987); Judith Butler, *Notes Towards a Performative Theory of Assembly* (London and Cambridge: Harvard University Press, 2015); Anna Lowenhaupt Tsing, *Friction: An Ethnography of Global Connection* (Princeton University Press, 2005), Kindle; Brian Massumi, *Politics of Affect* (Cambridge and Maldon: Polity Press, 2015); Astrida Niemanis, "Hydrofeminism or on Becoming a Body of Water," in *Undutiful Daughters: Mobilizing Future Concepts, Bodies and Subjectivities in Feminist Thought and Practice*, eds. Henriette Gunkel, Chrysanthi Nigianni and Fanny Söderbäck (New York: Palgrave Macmillan, 2012).

17 Tim Ingold, *The Life of Lines* (Oxon and New York: Routledge, 2015), 147.

18 Ingold, *Life*, 156.

19 Brian Massumi, *Parables of the Virtual: Movement, Affect, Sensation* (Durham and London: Duke University Press, 2002), 96, Kindle.

20 Massumi, *Parables*.

21 I have chosen to use the terms "in-between" and "interstitial" rather than "liminal" here because in many (though not all) contexts liminality describes a transitional state or site encountered within a process of moving between others which are more stable. The interstitial as I am discussing it does not necessarily suggest such stable beginning and end points, or regular process of movement between.

22 Significant for me in taking this approach is Maurice Merleau-Ponty's call for "plunging-into the world instead of surveying it", in *The Visible and the Invisible*, ed. Claude Lefort, trans. Alphonso Lingis (Evanston: Northwest University Press, 1968), 38–39.

23 Just a few of the many descriptors that will be encountered through this book include "gaps" in Tsing, *Friction*, "*différances*" in Jaques Derrida, *Margins of Philosophy*, trans. Alan Bass (Brighton: The Harvester Press, 1982), "chiasm" and "intertwining" in Merleau-Ponty, *Visible*, and "intra-action" in Karen Barad, *Meeting the Universe Halfway: Quantum Physics and the Entanglement of Matter and Meaning* (Durham and London: Duke University Press, 2007).

24 I am thinking in particular of Deleuze, Haraway, Derrida, Michel Serres, and Merleau-Ponty, whose terminologies for discussing in-betweens will be discussed through this book.

25 I will introduce this ontological approach more fully in Chapter 1.

26 In the Sydney region, this does not include the intertidal zones above the regular high tide mark. Saltmarsh is the plant community occupying these less frequently inundated zones in south-eastern Australia.

27 Alfredo Quarto, *The Mangrove Forest: Background Paper* (Gland: Ramsar Convention on Wetlands, 1997), www.ramsar.org/news/the-mangrove-forest-background-paper

28 Beth. A. Polidoro et al., "The Loss of Species: Mangrove Extinction Risk and Geographic Areas of Global Concern," *PLoS One* 5, no. 4 (2010), https://dx.doi.org/10.1371/journal.pone.0010095

29 Quarto, *Mangrove Forest*.

30 Norman Duke and Klaus Schmitt, "Mangroves: Unusual Forests at the Seas Edge," in *Tropical Forestry Handbook*, ed. Laslo Pancel and Michael Köhl (Berlin: Springer-Verlag, 2015); Kandasamy Kathiresan and Brian Lynn Bingham, "Biology of Mangroves and Mangrove Ecosystems," *Advances in Marine Biology* 40 (2001): 81–251.

31 Mark Huxham et al., "Mangroves and People: Local Ecosystem Services in a Changing Climate," in *Mangrove Ecosystems: A Global Biogeographic Perspective: Structure, Function, and Services*, ed. Victor H. Rivera-Monroy et al. (Cham: Springer International Publishing, 2017).

32 These names have been recorded with a variety of spellings, I have followed those listed by Val Attenbrow, "Clan Names Chart," *The Australian Museum*, 2018, accessed 1 May 2022, https://australian.museum/learn/cultures/atsi-collection/sydney/clan-names-chart/

33 Val Attenbrow, "Food from the Sea: Shellfish and Crustaceans," *The Australian Museum*, 2018, accessed 22 July 2020, https://australian.museum/learn/animals/crustaceans/food-from-the-sea-shellfish-crustaceans/; Paul Irish, *Aboriginal History Along the Cooks River* (Sydney: Cooks River Alliance, 2017), 11, 14.

34 Heather Goodall and Allison Cadzow, *Rivers and Resilience: Aboriginal People on Sydney's Georges River* (Sydney: University of New South Wales Press, 2009).

35 Daniel M. Alongi, "Mangrove Forests: Resilience, Protection from Tsunamis, and Responses to Global Climate Change," *Estuarine, Coastal and Shelf Science* 76, no. 1 (2008), https://dx.doi.org/10.1016/j.ecss.2007.08.024; Farid Dahdouh-Guebas et al., "How Effective Were Mangroves as a Defence against the Recent Tsunami?", *Current Biology* 15, no. 12 (2005), https://dx.doi.org/10.1016/j.cub.2005.06.008; Kandasamy Kathiresan and Narayanasamy Rajendran,

"Coastal Mangrove Forests Mitigated Tsunami," *Estuarine, Coastal and Shelf Science* 65, no. 3 (2005), https://dx.doi.org/10.1016/j.ecss.2005.06.022

36 Cheryl L. Doughty et al., "Mangrove Range Expansion Rapidly Increases Coastal Wetland Carbon Storage," *Estuaries and Coasts* 39, no. 2 (2016), https://dx.doi.org/10.1007/s12237-015-9993-8; Daniel Murdiyarso et al., *Carbon Storage in Mangrove and Peatland Ecosystems: A Preliminary Account from Plots in Indonesia* (Bogor: Center for International Forestry Research, 2009); Robert R. Twilley et al., "Productivity and Carbon Dynamics in Mangrove Wetlands," in *Mangrove Ecosystems: A Global Biogeographic Perspective: Structure, Function, and Services*, ed. Victor H. Rivera-Monroy et al. (Cham: Springer International Publishing, 2017).

37 Sarah E. Vaughn, "Disappearing Mangroves: The Epistemic Politics of Climate Adaptation in Guyana," *Cultural Anthropology* 32, no. 2 (2017): 261, https://dx.doi.org/https://doi.org/10.14506/ca32.2.07

38 Map 0.1 shows the locations of these two rivers relative to Sydney harbour and Botany Bay. Map 0.2 shows the locations of mangroves along the Cooks River. Map 0.3 shows the locations of mangroves along the lower reaches of the Georges River.

39 Ian Tyrrell, *River Dreams: The People and Landscape of the Cooks River* (Sydney: University of New South Wales Press, 2018), chapter 1, Kindle; Irish, *Aboriginal History*, 7–8.

40 Irish, *Aboriginal History*; Goodall and Cadzow, *Rivers and Resilience.*

41 According to Tyrrell, *River Dreams*, chapter 1, note 12, and Irish, *Aboriginal History*. note 18, the names used for the Cooks River by Wangal, Cadigal, and Gameygal people before 1788 have not been adequately recorded. Tyrrell reluctantly refers to the name *Goolay'yari*, meaning pelican landing, which has been adopted by councils and community groups for want of a better alternative, but Irish advises against this. I will therefore refer to this river as the Cooks River but note the use of this name is very heavily laden with the violence of invasion and colonial imposition.

42 Tyrrell, *River Dreams*, Introduction, paragraph 15.

43 Ibid.

44 Ibid., chapter 1, paragraph 28.

45 Ibid., chapter 4, paragraph 11.

46 Ibid., chapter 4, paragraph 7.

47 Ibid., chapter 3, paragraph 7; ibid., chapter 4, paragraph 2.

48 Bedigal and Dharawal names for the Georges River are no longer known, according to Goodall and Cadzow, *Rivers*, 29. These historians therefore do not refer to the river by any name other than the Georges River. Many popular sources claim that the historical indigenous name is Tucoerah.

49 Heather Goodall, Allison Cadzow, and Denis Byrne, "Mangroves, Garbage, Fishing: Bringing Everyday Ecology to Sydney's Industrial Georges River", *Transforming Cultures eJournal* 5, no. 1 (2010): 64, http://epress.lib.uts.edu.au/journals/TfC

50 Goodall and Cadzow, *Rivers and Resilience.*

51 Ibid., 29.

52 For examples and discussions of both these approaches, see Michelle Bastian, "Towards a More-Than-Human Participatory Research," in *Participatory Research in More-Than-Human Worlds*, ed. M. Bastian et al. (London: Routledge,

2017). For an example of Indigenous-led research that demonstrates methodologies beyond a human-centric focus, see Sarah Wright et al., "Telling Stories in, through and with Country: Engaging with Indigenous and More-Than-Human Methodologies at Bawaka, NE Australia", *Journal of Cultural Geography* 29, no. 1 (2012), https://dx.doi.org/10.1080/08873631.2012.646890

53 An overview of the emergence of these methodologies is provided in Eben S. Kirksey and Stefan Helmreich, "The Emergence of Multispecies Ethnography," *Cultural Anthropology* 25, no. 4 (2010), https://dx.doi.org/10.1111/j.1548-1360.2010.01069.x

54 Andrew Pickering, *The Mangle of Practice: Time, Agency, and Science* (Chicago: University of Chicago Press, 1995), 24.

55 Uli Beisel, Ann H. Kelly, and Noémi Tousignant, "Knowing Insects: Hosts, Vectors and Companions of Science," *Science as Culture* 22, no. 1 (2013): 8, https://dx.doi.org/10.1080/09505431.2013.776367

56 The framing of our entangled relationships as an experimental field is explored in Natalie Marr et al., "Sharing the Field: Reflections of More-Than-Human Field/work Encounters," *GeoHumanities* (2022), https//doi.org/10.1080/2373566X.2021.2016467

References

Alaimo, Stacey. *Exposed: Environmental Politics and Pleasures in Posthuman Times*. Minneapolis and London: University of Minnesota Press, 2016, Kindle.

Alongi, Daniel M. "Mangrove Forests: Resilience, Protection from Tsunamis, and Responses to Global Climate Change." *Estuarine, Coastal and Shelf Science* 76, no. 1 (2008): 1–13. https://dx.doi.org/10.1016/j.ecss.2007.08.024

Attenbrow, Val. "Clan Names Chart." The Australian Museum. Last modified 2018. Accessed 1 May 2022. https://australian.museum/learn/cultures/atsi-collection/sydney/clan-names-chart/

———. "Food from the Sea: Shellfish and Crustaceans." The Australian Museum. Last modified 2018. Accessed 22 July 2020. https://australian.museum/learn/animals/crustaceans/food-from-the-sea-shellfish-crustaceans/

Barad, Karen. *Meeting the Universe Halfway: Quantum Physics and the Entanglement of Matter and Meaning*. Durham and London: Duke University Press, 2007.

Bastian, Michelle. "Towards a More-Than-Human Participatory Research." In *Participatory Research in More-Than-Human Worlds*, edited by M. Bastian, N. Jones, O. Moore, et al., 19–37. London: Routledge, 2017.

Beck, Ulrich. *World at Risk*. Translated by Ciaran Cronin. Cambridge: Polity Press, 2009.

Beisel, Uli, Ann H. Kelly, and Noémi Tousignant. "Knowing Insects: Hosts, Vectors and Companions of Science." *Science as Culture* 22, no. 1 (2013): 1–15. https://dx.doi.org/10.1080/09505431.2013.776367

Berlant, Lauren. *Cruel Optimism*. Durham and London: Duke University Press, 2011, Kindle.

Butler, Judith. *Notes Towards a Performative Theory of Assembly. Harvard University Press*. London and Cambridge: Harvard University Press, 2015.

———. *Precarious Life: The Power of Mourning and Violence*. London and New York: Verso, 2004.

Clark, Nigel. *Inhuman Nature: Sociable Life on a Dynamic Planet. Theory, Culture and Society*. Nottingham: Sage Publications Ltd, 2011.

Connolly, William. *Facing the Planetary: Entangled Humanism and the Politics of Swarming*. Durham and London: Duke University Press, 2017, Kindle.

———. *Politics and Ambiguity: Rhetoric of the Human Sciences*. Madison: University of Wisconsin Press, 1987.

Coole, Diana. "The Inertia of Matter and the Generativity of Flesh." In *New Materialisms: Ontology, Agency, and Politics*, edited by Diana Coole and Samantha Frost, 92–115. Durham and London: Duke University Press, 2010.

Country, Bawaka, Sarah Wright, Sandie Suchet-Pearson, Kate Lloyd, et al. "Working with and Learning from Country: Decentring Human Authority." *Cultural Geographies* 22, no. 2 (2005): 269–83. https://dx.doi.org/10.1177/1474474014539248

Dahdouh-Guebas, Farid, Loku Jayatissa, Diana Di Nitto, Jared Bosire, et al. "How Effective Were Mangroves as a Defence against the Recent Tsunami?", *Current Biology* 15, no. 12 (2005): R443–7. https://dx.doi.org/10.1016/j.cub.2005.06.008

Deleuze, Gilles and Felix Guattari. *A Thousand Plateaus: Capitalism and Schizophrenia*. Translated by Brian Massumi. Minneapolis and London: University of Minnesota Press, 1987. Originally published as *Mille Plateaux: Capitalism et Schizophréne 2*, Les Éditions de Minuit, 1980.

Derrida, Jacques. *Margins of Philosophy*. Translated by Alan Bass. Brighton: The Harvester Press, 1982. Originally published as *Marges de la Philosophie*, Les Éditions de Minuit, 1972.

Doughty, Cheryl L., J. Adam Langley, Wayne S. Walker, Ilka C. Feller, et al. "Mangrove Range Expansion Rapidly Increases Coastal Wetland Carbon Storage." *Estuaries and Coasts* 39, no. 2 (2016): 385–96. https://dx.doi.org/10.1007/s12237-015-9993-8

Duke, Norman and Klaus Schmitt. "Mangroves: Unusual Forests at the Seas Edge." In *Tropical Forestry Handbook*, edited by Laslo Pancel and Michael Köhl, 1693–1724. Berlin: Springer-Verlag, 2015.

Gilson, Erinn Cunliffe. *The Ethics of Vulnerability: A Feminist Analysis of Social Life and Practice*. New York: Routledge, 2014, Kindle.

Goodall, Heather and Allison Cadzow. *Rivers and Resilience: Aboriginal People on Sydney's Georges River*. Sydney: University of New South Wales Press, 2009.

Goodall, Heather, Allison Cadzow, and Denis Byrne. "Mangroves, Garbage, Fishing: Bringing Everyday Ecology to Sydney's Industrial Georges River." *Transforming Cultures eJournal* 5, no. 1 (2010): 52–83. http://epress.lib.uts.edu.au/journals/TfC

Haraway, Donna. "A Cyborg Manifesto: Science, Technology, and Socialist-Feminism in the Late Twentieth Century." In *Simians, Cyborgs and Women: The Reinvention of Nature*, 149–181. New York: Routledge, 1991.

———. *Modest_Witness@Second_Millennium.Femaleman_Meets_Oncomouse: Feminism and Technoscience*. New York: Routledge, 1997.

———. *Staying with the Trouble: Making Kin in the Chthulucene*. Durham and London: Duke University Press, 2016.

Huxham, Mark, Amrit Dencer-Brown, Karen Diele, Kandasamy Kathiresan, et al. "Mangroves and People: Local Ecosystem Services in a Changing Climate." In *Mangrove Ecosystems: A Global Biogeographic Perspective: Structure, Function, and Services*, edited by Victor H. Rivera-Monroy, Shing Yip Lee, Erik Kristensen, et al., 245–274. Cham: Springer International Publishing, 2017.

Ingold, Tim. *The Life of Lines*. Oxon and New York: Routledge, 2015.

Irish, Paul. *Aboriginal History Along the Cooks River*. Sydney: Cooks River Alliance, 2017.

Kathiresan, Kandasamy and Brian Lynn Bingham. "Biology of Mangroves and Mangrove Ecosystems." *Advances In Marine Biology* 40 (2001): 81–251.

Kathiresan, Kandasamy and Narayanasamy Rajendran. "Coastal Mangrove Forests Mitigated Tsunami." *Estuarine, Coastal and Shelf Science* 65, no. 3 (2005): 601–06. https://dx.doi.org/10.1016/j.ecss.2005.06.022

Kirksey, Eben S. and Stefan Helmreich. "The Emergence of Multispecies Ethnography." *Cultural Anthropology* 25, no. 4 (2010): 545–76. https://dx.doi.org/10.1111/j.1548-1360.2010.01069.x

Kohn, Eduardo. *How Forests Think*. Berkeley and Los Angeles: University of California Press, 2013, Kindle.

Marr, Natalie, Mirjami Lantto, Maia Larsen, Kate Judith, Sage Brice, Jessica Phoenix, Catherine Oliver, Olivia Mason and Sarah Thomas. "Sharing the Field: Reflections of More-Than-Human Field/Work Encounters." *GeoHumanities* (2022). https://doi.org/10.1080/2373566X.2021.2016467

Massumi, Brian. *Parables of the Virtual: Movement, Affect, Sensation*. Durham and London: Duke University Press, 2002, Kindle.

———. *Politics of Affect*. Cambridge and Maldon: Polity Press, 2015.

Merleau-Ponty, Maurice. *The Visible and the Invisible*, edited by Claude Lefort. Translated by Alphonso Lingis. Evanston: Northwest University Press, 1968. Originally published as *Le Visible et L'Invisible*, Editions Gallimard, Paris, 1964.

Murdiyarso, Daniel, Daniel Donato, J. Boone Kauffman, Sofyan Kurnianto, et al. *Carbon Storage in Mangrove and Peatland Ecosystems: A Preliminary Account from Plots in Indonesia*. Bogor: Center for International Forestry Research, 2009.

Niemanis, Astrida. "Hydrofeminism or on Becoming a Body of Water". In *Undutiful Daughters: Mobilizing Future Concepts, Bodies and Subjectivities in Feminist Thought and Practice*, edited by Henriette Gunkel, Chrysanthi Nigianni, and Fanny Söderbäck, 85–99. New York: Palgrave Macmillan, 2012.

Pickering, Andrew. *The Mangle of Practice: Time, Agency, and Science*. Chicago: University of Chicago Press, 1995.

Plumwood, Val. *The Eye of the Crocodile*. Canberra: Australian National University E Press, 2012.

Polidoro, Beth. A., Kent. E. Carpenter, Lorna Collins, Norman C. Duke, et al. "The Loss of Species: Mangrove Extinction Risk and Geographic Areas of Global Concern." *PLoS One* 5, no. 4 (2010). https://dx.doi.org/10.1371/journal.pone.0010095

Povinelli, Elizabeth A. *Geontologies: A Requium to Late Liberalism*. Durham and London: Duke University Press, 2016, Kindle.

Quarto, Alfredo. *The Mangrove Forest: Background Paper*. Gland: Ramsar Convention on Wetlands, 1997. www.ramsar.org/news/the-mangrove-forest-background-paper

Rose, Deborah Bird. "Multispecies Knots of Ethical Time." *Environmental Philosophy* 9, no. 1 (2012): 127–40.

Shotwell, Alexis. *Against Purity: Living Ethically in Compromised Times*. Minneapolis and London: University of Minnesota Press, 2016, Kindle.

Tsing, Anna Lowenhaupt. *Friction: An Ethnography of Global Connection*. Princeton: Princeton University Press, 2005, Kindle.

———. *The Mushroom at the End of the World*. Princeton: Princeton University Press, 2015, Kindle.

Twilley, Robert R., Edward Castañeda-Moya, Victor H. Rivera-Monroy, and Andre Rovai. "Productivity and Carbon Dynamics in Mangrove Wetlands." In *Mangrove Ecosystems: A Global Biogeographic Perspective: Structure, Function, and Services*, edited by Victor H. Rivera-Monroy, Shing Yip Lee, Erik Kristensen, et al, 113–162. Cham: Springer International Publishing, 2017.

Tyrrell, Ian. *River Dreams: The People and Landscape of the Cooks River*. Sydney: University of New South Wales Press, 2018, Kindle.

Val Plumwood, *Feminism and the Mastery of Nature*, London and New York: Routledge, 1993, Kindle.

Vaughn, Sarah E. "Disappearing Mangroves: The Epistemic Politics of Climate Adaptation in Guyana." *Cultural Anthropology* 32, no. 2 (2017): 242–68. https://dx.doi.org/10.14506/ca32.2.07

Wright, Sarah, Kate Lloyd, Sandie Suchet-Pearson, Laklak Burarrwanga, et al. "Telling Stories in, through and with Country: Engaging with Indigenous and More-Than-Human Methodologies at Bawaka, NE Australia." *Journal of Cultural Geography* 29, no. 1 (2012): 39–60. https://dx.doi.org/10.1080/08873631.2012.646890

Yusoff, Kathryn. "Geologic Subjects: Nonhuman Origins, Geomorphic Aesthetics and the Art of Becoming Inhuman." *Cultural Geographies* 22, no. 3 (2014): 383–407. https://dx.doi.org/10.1177/1474474014545301

Part 1

Strange reflexivities

Folding in communicative tidal materialities[1]

This project adopts, performs, and explores a semiotic material approach to questions of being in-between. Part 1: "Strange reflexivities" filters the turns of semiotic material reflexivity through the tidal processes of the Georges River mangrove communities. The mangrove world lives by tidal change and regularity, incorporating these in a literal, embodied sense. Being tidal is a way of being that attends to tidal turns which change everything. The first section of this book considers ontological in-betweens within the tidal semiotics of mangroves and through the bodies of the oysters filter-feeding the tide-borne sediments along the rocky edges of the estuary. I follow the performance of the estuary as it is written and read through the responsive bodies of filter-feeding oysters.

The central problem tackled through this tidal encounter is the question posed by the contrast between the nested hierarchy of material semiotic levels theorised by biosemiotics, with the semiotic materialisms of Gilles Deleuze, Brian Massumi, Donna Haraway, Karen Barad, and Vinciane Despret, all of whom suggest models of semiotic materiality that are folded, fractal, and performative. Biosemiotics and the nested layered semiotics within the biological world which it has supported are the focus in Chapter 1. Folded ontologies are discussed in Chapter 2. Chapter 3 evaluates these approaches in relation to meaningfulness and tidal ecologies.

Note

1 Occasional passages within this section were published in Kate Judith, "How Mangroves Story: On Being a Filter Feeder," *Swamphen: A Journal of Cultural Ecology* 7 (2020), https://openjournals.library.sydney.edu.au/index.php/Swamphen/article/view/14366

DOI: 10.4324/9781003286493-2

1 The proposals of tides and the responses of oysters

Walking between roads and fences through streets in the early morning, birdsong grows louder and more varied as I descend towards Lime Kiln Bay on the Georges River in Sydney's southern suburbs. Leaving the streets, I follow a bush path along the edge of a small creek that widens into an estuary edged with mangroves. Entering along the boardwalk, I move into a darker world of gnarled trees, filtered light, and the half-glimpsed scuttling of thousands of crabs. As my attention settles into the mangroves, the sounds of lawn mowers and vehicles fade. So too does the expectation of a divide between solid and fluid. I become alert to mangrove movements, smells, light, and air quality. What matters changes. I immediately check the tide.

Visiting mangroves regularly, I become aware of tides. As I approach, the first question that automatically checks itself through my thought is: "where is the tide?" The question means more than it says, it is ontological: it is a matter of worlds. The mangrove is different worlds at different times during the tidal cycle. Each turn of the tide shifts the processes, relations, and behaviours of the place, re-mattering the estuary.

This chapter introduces semiotic materialism through the work that tidal rhythms do in proposing questions, or suggesting affordances, to which mangrove materialities become as response. Tidal flow asks questions of the shore: "how about this, and this, and this? What can you become with endless rhythm, sea bodies, and salt?" The shore becomes in response and asks it in return: "What can you become with nutrients, shallows, porous mud, and roots?" Tiny fish move in and find sanctuary during the high tides among the mangrove roots; mangrove propagules ride the waves as the tide retreats. The materialities of fish and mangroves are answers to these questions, and ask their questions in turn, back to the ocean and the shore. Oyster filter-feeding materialities become in response to tidal flows and rhythms, also working to shape these flows.

DOI: 10.4324/9781003286493-3

Tidal worlds and semiotic materiality

Tidal worlds

Tides regularly signal transformations in the Georges River mangal world. On the inward surge, the weight of the salty world laps its edges over the mangroves, bringing fragments of bone, shell, algae, and dead flesh. This is the awaited bounty of the tide, and many filter-feeding mouths are ready for its regular arrival. As the water rises, small fish move in among the pneumatophores with the incoming tide, quickly followed by larger fish and sometimes sharks who wait alongside. Danger follows safety as different but entangled opportunities open in succession. Birds who were feeding on the mudflats fly into trees to rest, groom, and socialise while cormorants and darters become more active, diving for fish along the edges of the tree line. The mangrove is now floored with a liquid surface reflecting the sky and the dark boughs that hang above it, the deeper channels glinting with the silver side turns of mullet and broken by their occasional airy leaps.

On the ebb tide, the mangrove breathes air rather than water and opens to land creatures. Gulls and ravens survey the mud to reach what they can before creatures slither off back into deeper channels. Humans and wading birds dig for worms and crustaceans in the surrounding mudflats. Semaphore crabs emerge from burrows, holding water around their gills for their water-based breathing.[1] The males begin signalling to each other across the mud, waving their claws up and around, as they graze on the slimy green algal and bacterial film, known as detritus, that covers the mud and stones. This film provides nourishment for many species; worms, shrimps, urchins, snails, and fish, and it helps bind the mud particles into the nutritious, spongy, mangal mudbody. Out from the mangal world, into the ocean, are washed wood, leaf, propagule, eggs, shells, oyster spawn, fish, plastics, fibres, and chemicals. When it is a mud place, the dark surface crawls with the scrambling and waving of crabs and the slow ramblings of whelks, and crackles eerily with the sounds of snapping shrimp.

Tidal swells occur as the Earth's fluids are pulled towards the moon's mass as the moon orbits about the Earth. Gravity, fluidity, and the movements of massive celestial bodies over long times make regular patterns of ebb and flow. That regularity establishes a temporality and spatiality that can potentially be interpreted, a tidal gap that can be occupied, that begs response. The rhythm poses as question, offers itself as affordance.[2] There might be a way to read that regularity into a body, a movement, or another rhythmic response. A rhythm can perhaps be lived within; chances can be measured against it. Something can, perhaps, make a dance or a life inside or against a tidal rhythm. In exploring rhythm, Elizabeth Grosz finds it "connects the most elementary and primary bodily structures of even the most simple organisms to the implacable movements of the universe itself".[3]

"Environment", the biosemiotician Wendy Wheeler writes, "is like a series of questions posed". She goes on to consider the interrelational quality of any question, suggesting that "the questions a system or an organism (non-human or human) chooses to answer will be those which, in its phenomenological epistemology, seem most urgently in need of a response".[4] Mangroves respond creatively and experimentally to the questions posed by tides. Further, the tidal swash of this relational story is not one-sided; mangroves return with their own questions back to the tide in the forms and capacities they build along the riverbanks. In considering my analysis in these terms, I am thinking of the estuary, the mangroves, and the tides as both material and semiotic, but not as though these are two ways of coding, one seeking to represent or creatively interpret the other, nor as though the meaning-making resides within the self-referentially closed semiotic material system, centred around the processes of human or another species' cognition. Here, in the tidal swash among the mangroves, meanings and materialities bring each other to bear together through affective, responsive intra-action.[5]

Semiotic materiality and its variants

Semiotic materialism is an ontological approach that finds meaningfulness and matter become together as a result of change. Interpreters, interpreted, relations, logics, differences, perceptual and cognitive processes, durations, and spatialities are emergent within the becoming meaningful of matter, within processes of mattering which fold semiotic and material together. Matter and meaning "emerge together in the world's process of becoming"[6] as "an ontological performance of the world in its ongoing articulation".[7] While no external driver or interpreter is required, its central interest in the processes of becoming meaningful distinguishes semiotic materialism from a strong materialism.[8] In the accounts of Gilles Deleuze, Donna Haraway, Brian Massumi, and Karen Barad, material reality makes itself meaningful through responsive, relational processes, and subjectivities emerge as material patterns develop self-reflexivity and some kind of integrity of pattern, while also interpreting and shaping. In these accounts, interstices are not problematic voids between opposing precepts but are necessary partners in the processes of becoming.

There are several divergent streams within the broad area I am describing as semiotic materialism or material semiotics (I interchange the wordings semiotic material and material semiotic, indicating the double-sidedness or intertwining of the semiotic and material elements of the concept),[9] some of which are visited through this book. Biosemiotics, which is a well-developed approach in its own right, considers semiotic materiality only within the context of biological life. Unlike many of the material semiotic approaches which extend beyond the biological sphere, it is not reliant upon a fundamental pre/concept located within all materiality.[10] Instead, it finds the processes of becoming meaningful within the mutual inter-responsiveness and co-becoming of organisms through the diverse processes of biological communications. Biosemiotics will be explored more fully later in this chapter. New materialism overlaps in many respects with semiotic materialism, to the extent that many writers can be situated within both approaches.[11] Like semiotic materialism this approach emphasises a non-anthropocentric materialist ontology that resists dualisms,[12] but while semiotic materialism pursues questions of the becoming-meaningful of matter, new materialism often takes a particular interest in questions of agency and subjectivity. In searching for agency without contemplating it as co-emergent with discursivity within semiosis, new materialism sometimes turns to a language of vitalism, exploring matter as agential or lively.[13]

Because semiotic materialism is a practice as much as it is an ontology, there are variations that approach from both angles. While new materialisms and biosemiotics predominantly read as ontological, Actor Network Theory (ANT) and material ecocriticism are positioned as methodological. The two angles, methodological and ontological, are ultimately interchangeable in this practice-based ontology. ANT, as developed by Bruno Latour and John Law, is a material semiotic variant identifiable by its particular mission to derive a set of methodological tools to support analysis in applying a material semiotic ontology to any analytical research question.[14] In doing so, it has enabled discussion of the many features that such an analysis is likely to have. ANT has been applied widely in studies of technologies,[15] scientific, agricultural and engineering practices,[16] and organisational structure.[17] The application of the ontology as a critical and/or creative approach, generally within an academic context, has been described as material ecocriticism.[18] Material ecocriticism "takes matter as a text, as a site of narrativity, a storied matter, a corporeal palimpsest in which stories are inscribed"[19] and engages then in critical interpretive practice with these stories/matter. Because of the interfoldedness of practice, meaning, and materiality involved here, semiotic materialism and material ecocriticism are interchangeable, if it is assumed (as I would argue it must be, given the non-anthropocentrism of the ontology) that material ecocritical practice is not confined to humans. In this book, I have used the terms semiotic materiality or material semiosis in preference to material ecocriticism, but this work is just as much a practice as an ontological exploration.

Some approaches, including some versions of material ecocriticism, adopt something like a weak form of semiotic materialism, limited to describing or analysing how human-centred storytelling or narrative practices, often literary texts, represent the agential materiality of the world. Serenella Iovino and Serpil Opperman explain that:

> Material ecocriticism proposes basically two ways of interpreting the agency of matter. The first one focuses on the way matter's (or nature's) nonhuman agentic capacities are described and represented

> in narrative texts (literary, cultural, visual); the second way focuses on matter's "narrative" power of creating configurations of meanings and substances, which enter with human lives into a field of co-emerging interactions.[20]

The first approach described in this quotation reflects this weak claim to materialism. By considering human cultural texts as producing *representations* of "nonhuman agentic capacities" which are then analysed by ecocriticism, this approach assumes the very conceptual differentiation between culture and nature that semiotic materialism otherwise overcomes. In the second way, the analyser of the artistic text, their analysis, the human artist, the cultural text they produce, and the matter that in the first approach was positioned as object represented are all already inter-affective within a semiosis.[21] By this account, cultural productions are never representations but always inter-foldings or diffractions of response, as is the analysis of them.

Along with many others, I regard semiotic materialism as not only a convincing approach to conceptualising meaningfulness, but also as practically applicable within relationships and decision-making, wherever these are occurring, in ways that can facilitate a non-anthropocentric reframing.[22] As Iovino and Oppermann explain, material ecocriticism "broadens and enhances the narrative potentialities of reality", expanding both what is regarded as agential and what counts as text.[23] Within a semiotic materiality, narrative potentialities and narrative strategies take on a broad relevance. They become applicable within relationships, in decision-making, and in considering social, economic, and political questions. The narrative potentialities of a semiotic materiality are particularly useful for the tasks of living responsibly within more-than-human communities and are practiced within many such communities. These narrative potentialities are the subject of the second section of this book, "Monstrous relations", and Chapter 6 in particular.

Biosemiotics

Biosemioticians are interested in all the processes through which information is expressed, perceived, communicated, and interpreted as meaningful within the biological.[24] This includes research into the semiotics of sensation and perception,[25] and the semiotics of the many and varied communications among the individuals of a species, but extends much further than that. The semiotics of communications among living things that are very different, such as between trees and insects, or orchids and bees,[26] is a notable area of discussion, as are the semiotics of cellular, evolutionary, and genetic processes.[27] While some of the biosemiotic literature is concerned with the details of semiotic processes between organisms, much explores the broader possibilities of a semiotic basis for the biological and entails an ontological commitment to a biological semiotic materiality. Wheeler, for example, writes, "nature is semiosic: made of signs, interpretations and meanings".[28] Within cells and organs the actions of the world are being read and translated into the complex narratives of bodies. Dynamic energetic inputs, and the emerging senses and semiotic frameworks that are interpreting them inter-affect each other and communicate each other into discernible patterns. Mattering and meaning-making are being worked within processes that are both material and semiotic. The three biosemioticians whose work I will predominantly be referring to are Jesper Hoffmeyer, Terrence Deacon, and Wendy Wheeler.

Biosemiotic work approaches biological processes as semiotic, exploring "what, precisely, such ineliminable terms as 'meaning' and 'significance' might refer to in the context of living, complex adaptive systems".[29] The field is searching for terms which can probe the processes through which complex meanings and relations become, so draws from both biological and linguistic disciplinary traditions without privileging or subsuming one over the other. In this context words like "decision", "agent", "choice", "story", "reading", "writing", "mind", and "self" open to their non-anthropocentric, material processual potentials,[30] while words like "body", "cell", or "photosynthesis" also evoke semiotic implications. This aligns with the semiotic materialist position of thinkers such as Barad, Vicki Kirby, and Jane Bennett who demonstrate few reservations in employing such encompassing language usages.[31] The meanings of concepts like "thinking", "intelligence", and "decision" open to make sense within each particular, situated context, rather than always assuming a human-centred context. For example, if I can hold open the idea of "deciding" to allow that there is a mangrove way of deciding, which will be particular to the events occurring within mangroves, I can make space in my sense of the world for mangrove agency. A mangrove way of deciding is coded within its biology and also depends upon its circumstances and relationships. When a mangrove propagule decides to send down roots, for example, depends upon the way its complex, adapted chemistry and biology respond to the availability and type of substrate, the amount of turbulence, and a range of other environmental conditions. These factors are intra-affecting, so the decision-making process is not prescribed by them in a simple way.

Environments are communicative and full of signs relative to perspectives. As Jacob von Uexküll illustrated using the example of an oak tree, one thing can signify very different meanings for different creatures.[32] For a fox the oak is shelter, for an owl it is a snug resting place, for a beetle it is food, and so on. In addition, "shared interpretive universes or motifs" or "ecological discourse structures"[33] can develop among organisms. Hoffmeyer uses the example of a hare and a fox who both understand – each in their own way – what the hare "means" when it stands up and shows itself to the watching fox. According to Hoffmeyer, the hare does this to show the fox that the fox has been seen, and can be outrun if needed, making the pursuit unnecessary. Both apparently read this as a sign that no hunting or fleeing will be needed. In this way, fox and hare share an interpretive motif. What counts as a sign within biosemiotics depends upon the communicative context. It depends upon what is being interpreted, how, and by whom or what. Hoffmeyer explains; "[T]he sign doesn't exist materially, it happens. The sign is an event, a process guided by the ingrained logic hidden in the receptive system".[34] Deacon places the semiotic action within the interpretive process, removing the independence and separability of subject, object, and sign. Interpretation is entangled within the biological process throughout. Who or what is interpretant, who or what is interpreted, and who or what is sign are contextual.[35]

Biosemiotics shares all of the above features with the semiotic materialities of Deleuze, Haraway, Barad, and Massumi. These two broad approaches to semiotic materialism are very alike. They differ, however, in two important ways. First, biosemiotics limits the scope of semiotic materiality to biological processes, whereas the other expands it to encompass all meaningful materialities. Second, biosemiotic narratives trace semiotic material complexity as increasing across thresholds, from which new interpretive orders emerge. In the biosemiotic literature I have read, this is not an additional feature imposed upon, but potentially separable from the key processes. It is central to the biosemiotic mechanism. In these accounts, it is only because there are jumps in abstractions that biological materiality becomes semiotic. The narrative of thresholds across layers becomes increasingly problematic when it is connected with an allocation of value within a story of progress. When biosemiotic accounts separate and classify biosemiotic processes into hierarchical levels, associating greater biosemiotic value with the most abstract level (which is generally identified as that of human language and cognition) as most masterful, they continue to imagine a natural order culminating in the exceptional human.

Nested layers or levels

A hierarchy of layers

A nested hierarchy of layers with human consciousness at its pinnacle remains a widespread feature of biosemiotic models. Hoffmeyer, for example, elucidates a hierarchy of levels of increasing semiotic competency that step up from the cellular, to the unicellular organism, to the multicellular body, and the ecosystem, with increasing levels also noted through the development of brains.[36] Timothy Ryan Day summarises the general story:

> From bacteria transporting DNA, to central nervous systems, to the abstract thought that leads to human language and now [...from] the leap of human language into the even further abstractions of AI,

material has a proven history of exceeding its limits and creating ever more discursive selves.[37]

Wheeler's summary of her position demonstrates the emergent nature of each level; "evolution proceeds by the development of layers (or levels) of increased complexity, with emergent qualities which produce hierarchical and supervenient levels (or layers) of more (communicational, semiotic) complexified organisation".[38] Like Hoffmeyer and Deacon,[39] she claims the human employment of symbolic logic constitutes a novel logical order or level.

This feature can be traced back to biosemiotics' two historical and theoretical antecedents; systems ecology and Peircean semiotics. A biological sphere organised as a nested hierarchy of layers of organisational complexity, cell, individual, species, community, ecosystem, biome, and biosphere, developed as a heuristic through the mid-twentieth century within the growing discipline of systems ecology. This young discipline drew heavily from systems theory, as evidenced by the influential concept of the ecosystem.[40] Systems theory itself first developed within an engineering context concerned with the flows of energy across gradients within a bounded system.[41] In its early decades, ecosystem science took a similar interest in such gradients and transfers. Food pyramids or chains, for example, organised as trophic levels were based on the transfer of energy as organisms consume each other, with photosynthesisers at the bottom up to apex predators at the top, with decomposers performing a closure of the circle by releasing nutrients to be again available for plants and algae. According to Raymond Lindeman's 1942 model, across each step approximately 90% of the energy stored within the lower level is lost to the environment outside the trophic pyramid through processes such as respiration and incomplete digestion, the remaining 10% being transformed into a new organismic form through eating and digesting.[42] Because of this, in a trophic pyramid, higher levels appear to achieve something like a higher scale of energy concentration. It can seem as though some kind of status directionality across scales is at play, as though there is a sense in which the lower level finds its purpose in providing the energy for the higher energy scale.

Ecological and biological theory has moved on from overly simplistic conceptualisations of food web levels and evolutionary trees with simple hierarchies of levels. There is now widespread recognition that the movements of both energy and genetic information are more complex than previously thought. The actions of viruses and bacteria keep most of the world's nutrients looping around within the ocean's microbial life, never "going up the food chain" or evolving into something more complex.[43] Microbes and viruses transfer the energetic and genetic building blocks of biological life between and within bodies in ways that scramble and complicate emerging orders,[44] and local contexts affect genetic expression and behaviours more than previously expected.[45] Narratives of progress through hierarchical steps can be exchanged for complex twisted plots.

Peircean semiotics: habit

The other theoretical ground for the hierarchy of levels within biosemiotics is Peircean semiotics. Charles Sanders Peirce developed a radically different approach to semiotics from that most widely applied within the context of human language and has been influential for many thinkers who have sought a model to explain decision-making, agency, and cognition among the living and non-living forms of the world that does not ultimately require a step involving human rationality. Where mainstream semiotic theory[46] assumes a symbolic linguistic realm independent from and representative of the material world and has until relatively recently been applied almost exclusively to human language use,[47] Peircean semiotics has a broader more general scope that has allowed it to be applied much more widely.

The basis of Peircean semiotics is habit, which establishes in response to regularity.[48] The process involves a particular kind of logic that Peirce calls abduction, which he describes as "the process of forming an explanatory hypothesis" and that he proposes underpins life.[49] Massumi describes abduction from an embodied perspective as "thought that is still couched in bodily feeling, that is still fully bound up with unfolding sensation as it goes into action but before it has been able to articulate itself in conscious reflection and guarded language".[50] Peirce suggests life comes into its forms and relationships via an abductive rather than reasoned logic through responding to metaphoric insights. A fish that is hungry and eats a mangrove leaf is reading it as a meal; a fish that sees a heron and hides under a leaf is reading it as shelter. Peircean abduction would explain these moves as made through an event of contiguity between sensual data and bodily desire or need. If these actions become habituated, the fish would no longer need the impetus of hunger or fear to read the leaf in the way most useful for its needs – she habitually moves under a leaf immediately she sees a heron. At no stage is there a need for recourse to reasoned logic. Habits develop as abductions become organised and simplified into semiotic categories, not necessarily consciously or linguistically. The oyster's body effectively categorises the edible from the inedible using complex fluid dynamics created by its shell and gill shape and the particular spiralling movements of hair-like cilia, bodily behaviours of semiotic classification evolved over generations into an always-responsive species form or material habit (the physiological processes of oyster filter-feeding will be explored further shortly).[51] Because of this foundation in the establishment of habit, semiotic material approaches that rely upon Peircean principles often emphasise the importance of regularity. Kohn, for example, expresses a clearly Peircean approach when he writes, "it is only because the world has some semblance of regularity that it can be represented. Signs are habits about habits".[52]

A Peircean semiotics based upon habits postulates scales of efficiency as life becomes complex. Habituation means groups of localised and specific responses can be subsumed into one habit, eliminating the need to invoke the effort of abductive insight for each similar encounter. Marcel Danesi describes the scalability enabled by a Peircean semiotic approach,

> Once an insight is attained, it becomes useful to "routinize" it, so that a host of related problems can be solved as a matter of habit, with limited time-consuming mental effort. … This is the rationale behind all organized knowledge systems.[53]

Abductive habit is an abstracting move. It bundles many different things together so they can then be considered as one (not necessarily consciously – this is all in the context of a material semiotic), and it creates the position from which this occurs. A landscape of hierarchical levels of greater abstraction can occur as bundles themselves become bundled. Along with the methodological explanation for material semiosis, many thinkers who turn towards Peircean semiotics for its more-than-human semiotic logic also adopted a semiotic material landscape comprising nested scales or levels of semiotic abstraction or efficiency. The layering effect of a Peircean approach is emphasised by its influential three-tiered framework.

Peircean three-tiered framework

Peirce developed a tripartite framework to explain in detail the processes of semiosis within his model. His terminology identifies three key elements. "Firstness" is the occurrence of some kind of suggestion or feeling that there may be a pattern; it is the vague scent of a pattern.[54] "Secondness" is the pattern habituated so it is experienced as an "otherness" or an object,[55] and "thirdness" is the process mediating between these two, it is the tendency to take habits, the processes of becoming generalised.[56] It is through thirdness that there is an interpretant, a semiotic performer, or performative process. Thirdness effects a reflexive move within materiality, where a difference becomes folded into an otherness and an interpretant. Peirce locates meaningfulness with thirdness; "the word **means** is almost an exact synonym to the word **third.** It certainly involves Thirdness".[57] Any sign[58] includes these three within the process of its becoming. Importantly, within this model, semiosis is a happening rather than something inherent to already established relations, and meaningfulness only continues if the reflexive relationship between the three aspects remains actively performed.[59] Because of this basic mechanism, Peircean semiotics is both steadfastly material and maintains a reflexive move that builds-in a jump in abstractive levels.

Peircean semiotics can be and has been employed reductively as though its three core elements were separately identifiable and graspable. A reductive Peircean account might parse filtering into three components following this model: the unfiltered (as the first, full of potential for becoming meaningful), the fully separated (the second, where difference is distinguished), and the filter performing the separation (the third, which provides the criteria and context for meaning). Such an account would treat the filtered and unfiltered as distinct from each other and distinguished by their differences. The third rearranges the world through filter paper, photographic filter, or the body of an oyster, writing a layered coherence into the otherwise hoveringly potential. In doing so, it provides the criteria that make the difference meaningful. Together the three elements enable that meaningfulness. Peircean semiotics allows meaning to be considered as inherent within the materiality of the world, rather than being a feature of human language, but it comes associated with a logic that has often been used to support stable layering. Biosemiotics that has adopted Peircean semiotics for the former purpose has frequently also adopted the latter.

This limitation has been identified by some commentators. Vefa Karatay, Yagmur Denizhan, and Mehmet Ozansoy, for example, write: "Obviously far ahead of his time, very inspiring and without doubt processual in nature, Peircean semiotics nevertheless carries rather static elements that seem to be remnants of hylemorphism and atomism".[60] They also suggest Peirce's model of "chains of triads" is not sufficiently flexible to account for the dynamics of a universal and fundamental semiosis. In their view, "even a single triad should be seen not as an abstract logical schema but as a dynamic ontogenetic process – no matter how well-established or 'inveterate' it may be".[61] This is similar to the position Vicki Kirby takes in her critique of the tripartite model of subject, object, and intermediating method.[62] She argues the subject is necessarily more involved than this account, which removes it to an "external and precedent" position, allows. Kirby turns to Barad's account of quantum entanglement for illustration of an "involved transubstantiation of possibles whose differences remain entangled",[63] arguing that entities emerge entangled, not separable, that to exist is already to be entangled.[64] This is also well illustrated by the oyster, whose relationship with the tidal current and the geobiome of the mudbank does not deny the oyster's existence but shows it as necessarily entangled. Change the current, and the oyster and the mudbank change as well. The Peircean three-step process, if it is to avoid hierarchical layers that tend towards anthropocentrism, might need to look and feel more like the oyster body becoming in response to the tidal currents and the muddy mangrove world taking shape from the historied and situated sorting processes of the oysters.

Filter-feeding

Filter-feeding: flow and constraint

Tidal flow offers risk and opportunity for a thing that is loosened from its connections. It affords a means of dispersal for many creatures and is especially useful for the dispersals of gonads, spawn, propagules, and seeds of mangrove dwellers. Oysters spawn when they perceive conditions such as salinity, temperature, and the presence of phytoplankton are favourable. The first oyster releasing its gonads signals the others nearby to do so also, aligning the reef in a reproductive chorus.[65] Fertilised eggs drift in the water current growing into free-floating larvae who then grow a foot enabling them to crawl and search for a suitable substrate. They chance their futures upon the flow's apparent inevitability and reliability, into hopeful but unpredictable futures. Holding on tight, resisting the force of the flow, and utilising its energy is another responsive material reply to regular tidal flows, and many tidal creatures use both strategies. When they hold fast, the oyster's relationship with the tide turns around. Instead of working with the tide for its own movement, it risks its question to the tide as a deliverer of sustenance, becoming filter-feeder.

Many mangrove organisms filter food from the flowing current. Some of the most visible filter-feeders of the Cooks and Georges River mangroves are the many Sydney rock oysters (*Saccostrea glomerata*), which hug the rocks or walls of embankments. Before the British invasion, expansive oyster reefs thrived further out in these estuaries and in Botany Bay and provided food for people for thousands of years. These were extensively mined for food and lime in the early years of Sydney's growth, and then killed completely when a virulent parasitic mudworm species was accidentally imported on oyster spawn introduced from New Zealand.[66] Large sub tidal shellfish reefs like these were widespread along Australia's east coast estuaries but are now the nation's most threatened marine habitats, considered critically endangered with over 90% destroyed.[67] Australian selfish reefs are not alone. Globally, around 85% of these ecosystems have been destroyed, making them "the most imperilled marine ecosystem on earth".[68]

Far from the passive life forms they are often imagined to be, filter-feeders employ complex physiological mechanisms to enable particles from the water current to find their way to their digestive systems. Bivalves such as oysters and mussels create complex spiralling currents using the movements of hair-like structures called cilia that work with a rowing action, diverting the surrounding water currents towards their environments.[69] The cilia move in an "oscillating unsteady 3-dimensional pattern" to further transport particles deftly through the animal's folded gills towards mucus near their mouths using fluid mechanics.[70] Not all particles are digestible, so filter-feeders also require some kind of sorting mechanism. Oysters sort using complex fluid dynamics, creating diverting flows through their bodies. The shape and quality of the current created through the body by the cilia, in combination with the folded shape of the gills, send particles of different densities and even different surface qualities in different directions, either to a mucus net to be excreted as pseudofaeces, or to the mouth.[71] An oyster's fluid mechanics is refined enough to select the potential food from similar sized and shaped non-digestible particles.[72]

Experimentally, over evolutionary time, bivalve bodies have become in response to the characteristics of viscous flow. Beyond this, their bodies have become engineers of flow, able to work experimentally with it, to design flow so the complex three-dimensional movements of cilia direct particles towards mouths or sweep them towards expulsion as pseudofaeces. Cilia, mucus, and the shapes and smoothness of shells and bodies are attuned to viscous tidal fluidity. To use the language of this chapter's title, they are taking the energy of a tidal current as a question, answering it with their capabilities, asking back to the current what it is capable of.[73] Flow and bodies respond to each other's capabilities, modifying over both very long and very short time durations.

The filter-feeding processes of these bivalves have several other profound environmental impacts. Their excreted material accumulates as the geobiome of the muddy substrate of the estuary. Different species and genera of bivalves excrete different microorganisms with their faeces and pseudofaeces,[74] producing mud with different biochemical capabilities, writing the microbial character of the estuarine mud. The ancient oyster reefs of Botany Bay and its tributaries cleared the water through their filter-feeding,[75] supporting the health of seagrass meadows.[76] Without shellfish, however, the water remains more turbid and less optimal for seagrass ecology. Bivalve filter-feeders also remove nitrogen, trace metals, pesticides, and other pollutants from the water and deposit the nitrate rich waste into the mud from their faeces or pseudofaeces, where bacteria process it into nitrogen gas.[77] By doing this, oyster reefs significantly reduce nitrogen passing into open waters.[78] Juvenile oysters frequently fasten onto the shells of adult oysters, so over time large reef systems of oysters can grow, and these can provide excellent niches for many other species, as they once did in the Georges River estuary and Botany Bay. In summary, oyster filter-feeding contributes to the structure, habitats, food resources, chemistry, and biochemistry of an estuarine ecosystem.

Disturbance and semiotics askew

Rhythms are regular but not simple as they move across the mangrove. Tides are complicated by the shape and slope of coasts and riverbanks, and the sun's gravitational forces complicate those of the moon, tugging the spring tides higher, toning down the ebb tides when oysters sleep. Additional rhythms are interfolded within the tidal. Wind-generated waves lick their energy against the mud and root mass of the mangrove body. Some waves are tiny, moving small particles, helping plankton wash into a space between mud grains. Others are huge, destructive, landscape-altering, mangrove-destroying. The rhythm of night and day transforms the mangrove too, as diurnal and nocturnal creatures and processes awaken and repose. Annual seasonal rhythms are layered into the mix, and every animal and plant changes to the rhythms of their own body's temporalities. Floods, rain, and drought also diffract within these rhythms, disturbing them in both predictable and unpredictable ways. Estuarine rhythms hold out the tantalising suggestion that their regularities can be read and interpreted by life's opportunistic, experimental processes, but the task remains challenging, the rhythms are complexly interrelated, and many disruptions are unpredictable. In ever-diminishing worlds within worlds, gaps of constancy are troubled by variance. The ever-presence of unpredictable disturbance within every apparent regularity sets a lean or veer into the semiotics of materiality. When meaning is made on the run, in changing circumstances, habits are annoyed before they establish, the processes of becoming meaningful are set askew, and semiotic layers of abstraction fail to settle. Peircean semiotics emphasises regularity to explain how meaning becomes possible, but this is arguably at the expense of consideration of the ubiquity of disturbance.

Approaches to semiotic materiality which emphasise the importance of continual change and reject the necessity of nested layers of abstraction have drawn upon a variety of theory, including Peircean semiotics. Deleuze and Massumi are influenced by the process-based philosophy of Henri Bergson. Barad's work begins from observations within quantum mechanics particularly drawing upon the work of Niels Bohr. William Connolly, Bennett, Latour, Vinciane Despret, and Haraway are among those working largely within science and technology studies. Haraway has, in turn, been influential for many feminist new materialists, and Kirby finds her way into material semiotics through Derridean post-structuralism. Timo Maran, reflecting upon what distinguishes these thinkers (whom he describes as material ecocritics) from the biosemioticians, writes: "there is a crucial difference between material ecocriticism and biosemiotics; whereas the former has taken a critical approach to human social and cultural processes, the latter has not".[79] For Maran the crucial difference is not within their onto-epistemological models but in the extent to which the work engages critically with human social and cultural processes. While this distinction is apparent, I would add that there are significant onto-epistemological differences between the two broad approaches, centred upon whether a semiotic landscape of nested levels or fractal folds is imagined. The next chapter considers folded semiotic material approaches.

Tidal turns bring material, regular, shifts in the world – all movement pulls one way, there is a pause, a turn, and movement pulls the other way. Many living and non-living things and processes physically make a turn or associate with different other things as the tide turns. The wading birds change from resting to eating, the pneumatophores change from interfacing salt to interfacing air, the crayfish from hunting to hiding, the small fish from wary to relaxed. Tidal turns also layer the earth in sediment, settled by flocculation into mudbanks, eventually into layers of sedimentary mudstones and shales, in many places layered also with the fossilised bodies of ancient plant and animal inhabitants, or metamorphosed into the hardened layers of slates. Layers sediment into strata when considered from afar, temporally and spatially or across scales of semiotic abstraction. Narratives of emergent layers reflect a perspective that locates meaning with a position that can stand outside, and with stasis, continuity, and regularity.

Entering into the thick between, animals, currents, plants are actively figuring their spatialities and temporalities on the move. In Chapter 2, I turn to consider a range of concepts derived from the work of a diverse group of thinkers, concepts that support material semiotic ontologies which seek to value change equally with the becoming of coherence, and to maintain the position of interpretants as always within and becoming with the semiotic process.

Notes

1 Norman Duke and Klaus Schmitt, "Mangroves: Unusual Forests at the Seas Edge," in *Tropical Forestry Handbook* (Berlin: Springer-Verlag, 2015), 1713–1714.

2 For James Gibson, who introduced the concept: "The affordances of the environment are what it offers the animal, what it provides or furnishes, either for good or ill", in *The Ecological Approach to Visual Perception*, Classic ed., (New York: Taylor and Francis, 2015), 119. This might suggest a discourse of exchange where animal perception takes from an environment that offers, which is not the sense I intend here. Through this book I take "affordance" to mean "the responses enabled for an organism by a situation", which I imagine as combining Chemero's relational and Pickering's semiotic versions of the concept, as described in Anthony Chemero, "An Outline of a Theory of Affordance," *Ecological Psychology* 15, no. 2 (2013); John Pickering, "Affordances Are Signs," *Triple C: Cognition, Communication and Co-operation* 5, no. 2 (2007), www.triple-c.at/index.php/tripleC/article/view/59/61

3 Elizabeth Grosz, *Chaos, Territory, Art: Deleuze and the Framing of the Earth* (New York: Columbia University Press, 2008), 19, Kindle.

4 Wendy Wheeler, "Gregory Bateson and Biosemiotics: Transcendence and Animism in the 21st Century," *Green Letters* 13, no. 1 (2010): 37, https://dx.doi.org/10.1080/14688417.2010.10589069

5 "Intra-action" is Barad's term, coined to describe the processes through which matter writes itself into its discursive forms, "the mutual constitution of entangled agencies" in Karen Barad. *Meeting the Universe Halfway: Quantum Physics and the Entanglement of Matter and Meaning* (Durham and London: Duke University Press, 2007), 85. Barad contrasts the term with "inter-action", which, she claims, retains the sense of separate agencies operating across a gap.

6 Serenella Iovino and Serpil Oppermann, "Theorizing Material Ecocriticism: A Diptych," *ISLE: Interdisciplinary Studies in Literature and Environment* 19, no. 3 (2012): 435, https://dx.doi.org/10.1093/isle/iss087
7 Barad, *Meeting*, 149.
8 Strong materialism finds explanation only within material processes and rejects explanations requiring any transcendence. Semiotic materialisms, in contrast, often involve transcendent semiotic moves *within* their material accounts. This will be explained more fully and discussed further in Chapter 2.
9 I note that Donna Haraway calls her approach "material semiosis and semiotic materiality" using the two wordings together as interconnected, in Donna Haraway, *When Species Meet* (Minneapolis: University of Minnesota Press, 2008), 136.
10 Some examples of these are intra-action (Barad), affect (Massumi), différance (Derrida), and diffraction (Haraway and Barad). Some of these will be discussed later within various chapters.
11 My understanding of new materialisms aligns with that provided within Diana Coole and Samantha Frost, eds., *New Materialisms: Ontology, Agency and Politics* (Durham and London: Duke University Press, 2010), Kindle.
12 Iris van der Tuin and Rick Dolphijn, "The Transversality of New Materialism," *Women: A Cultural Review* 21, no. 2 (2010): 156, https://dx.doi.org/10.1080/09574042.2010.488377
13 A key text discussing this is Jane Bennett, *Vibrant Matter: A Political Ecology of Things* (Durham and London: Duke University Press, 2010).
14 John Law, for example, describes Actor Network Theory as "a disparate family of material-semiotic tools, sensibilities, and methods of analysis that treat everything in the social and natural worlds as a continuously generated effect of the webs of relations within which they are located. It assumes that nothing has reality or form outside the enactment of those relations", in "Actor Network Theory and Material Semiotics," in *The New Blackwell Companion to Social Theory*, ed. B. S. Turner (Oxford: Blackwell, 2009), 141.
15 For example: Harun Kaygan, "Material Semiotics of Form Giving: The Case of the Electric Turkish Coffee Pot," *Design Issues* 32, no. 2 (2016), https://dx.doi.org/10.1162/DESI_a_00384
16 For example: Mara Miele, "The Making of the Brave Sheep or … the Laboratory as the Unlikely Space of Attunement to Animal Emotions," *GeoHumanities* 2, no. 1 (2016), https://dx.doi.org/10.1080/2373566x.2016.1167617
17 For example: Michel Callon, "Revisiting Marketization: From Interface-Markets to Market-Agencements," *Consumption Markets & Culture: Agencing Markets* 19, no. 1 (2016), https://dx.doi.org/10.1080/10253866.2015.1067002
18 Serenella Iovino and Serpil Oppermann, "Material Ecocriticism: Materiality, Agency and Models of Narrativity," *Ecozon@* 3, no. 1 (2012): 75–91.
19 Iovino and Oppermann, "Material Ecocriticism", 84.
20 Ibid., 79–80. Iovino and Oppermann are ambiguous on this issue. In several places they explicitly state the necessity of a human within meaning-making relations, whereas in other passages they explicitly state this is not required.
21 This point is emphasised by Barad throughout her work, for example: "We do not obtain knowledge by standing outside of the world; we know because 'we' are of the world. We are part of the world in its differential becoming", in "Posthumanist Performativity: Toward an Understanding of How Matter

Comes to Matter," *Signs: Journal of Women in Culture and Society* 28, no. 3 (2003): 147.

22 Other writers who have also made this point include Haraway, Latour, Barad, Alaimo, Iovino, and Oppermann.

23 Iovino and Oppermann, "Theorizing Material Ecocriticism", 459.

24 The International Society for Biosemiotic Studies website features this definition of the discipline: "an interdisciplinary research agenda investigating the myriad forms of communication and signification found in and between living systems", in International Society for Biosemiotic Studies, "What Is Biosemiotics?", accessed 26 July 2020, www.biosemiotics.org/what-is/

25 Jesper Hoffmeyer, "Semiotic Scaffolding: A Unitary Principle Gluing Life and Culture Together," *Green Letters* 19, no. 3 (2015): 246, https://dx.doi.org/10.1080/14688417.2015.1058175

26 Several examples of research exploring plant/insect communications are provided in Carla Hustak and Natasha Myers, "Involutionary Momentum: Affective Ecologies and the Sciences of Plant/Insect Encounters," *Differences: A Journal of Femininst Cultural Studies* 25, no. 5 (2012), https://dx.doi.org/10.1215/10407391-1892907

27 Jesper Hoffmeyer includes discussion of inter- and intra-cellular and genetic biosemiotics, in "Biosemiotics: Towards a New Synthesis in Biology," *European Journal for Semiotic Studies* 9, no. 2 (1997).

28 Wendy Wheeler, "A Connoisseur of Magical Coincidences: Chance, Creativity and Poiesis from a Biosemiotic Perspective," *Biosemiotics* 7, no. 3 (2014): 375, https://dx.doi.org/10.1007/s12304-014-9218-8

29 International Society for Biosemiotic Studies, "What is Biosemiotics?".

30 See for example: Terrence W. Deacon, James W. Haag, and Jay Ogilvy, "The Emergence of Self," *In Search of Self: Interdisciplinary Perspectives on Personhood*, eds. J. Wentzel van Huyssteen and Erik P Wiebe (Grand Rapids: Wm. B. Eerdmans Publishing Co, 2011); and Eduardo Kohn, *How Forests Think* (Berkeley and Los Angeles: University of California Press, 2013), 16, Kindle.

31 Vicki Kirby, for example, writes: "life, here biology, is argumentative, left and right-leaning, aggressive, ruthless, generous, self-sacrificing, caring [...] and in a constant state of transition and metamorphosis because its own self reflections are volatile", in "Autoimmunity: The Political State of Nature," *Parallax* 23, no. 1 (2017): 56, https://dx.doi.org/10.1080/13534645.2016.1261661. Barad describes how matter "feels, converses, suffers, desires, yearns and remembers" cited in Rick Dolphijn and Iris van der Tuin, *New Materialism: Interviews & Cartographies* (Ann Arbour: Open Humanities Press, 2012), 59. http://hdl.handle.net/2027/spo.11515701.0001.001

32 Jacob von Uexküll, "A Stroll through the Worlds of Animals and Men: A Picture Book of Invisible Worlds," in *Instinctive Behaviour: The Development of a Modern Concept*, ed. and trans. Claire H. Schiller (New York: International Universities Press, Inc., 1934), 73–78.

33 Hoffmeyer, "Biosemiotics".

34 Hoffmeyer, "Semiotic Scaffolding", 245.

35 Terrence W. Deacon, "Steps to a Science of Biosemiotics", *Green Letters* 19, no. 3 (2015), https://dx.doi.org/10.1080/14688417.2015.1072948, 301.

36 Hoffmeyer, "Biosemiotics"; Hoffmeyer, "Semiotic Scaffolding", 247.

37 Timothy Ryan Day, "With Parted Eye: A Midsummer Night's Dream, Richard Powers' Orfeo, and Biosemiotics," *Green Letters* 22, no. 3 (2018): 253, https://dx.doi.org/10.1080/14688417.2018.1512418.

38 Wheeler, "Bateson", 38.

39 Deacon, Haag, and Ogilvy, "The Emergence of Self," 15–18.

40 Howard T. Odum and Eugene P. Odum, *The Fundamentals of Ecology* (Philadelphia: Saunders, 1953).

41 Prior to Ludwig von Bertalanffy's *General System Theory: Foundations, Development, Applications* (London: Allen Lane The Penguin Press, 1971), which introduced and explored the concept of open systems, systems thinking had been associated with closed systems in the context of classical physics and its applications within engineering.

42 Raymond L. Lindeman, "The Trophic-Dynamic Aspect of Ecology," *Ecology* 23 (1942), https://dx.doi.org/10.2307/1930126

43 Maureen O'Malley, "The Ecological Virus," *Studies in History and Philosophy of Biological and Biomedical Sciences* 59 (2018), https://dx.doi.org/10.1016/j.shpsc.2016.02.012

44 An example of a research area that has contributed towards furthering understanding of this complexity is the endosymbiotic theory of eukaryotic cells developed by Lynn Margulis. See, for example: Lynn Margulis, "Genetic and Evolutionary Consequences of Symbiosis," *Experimental Parasitology* 39, no. 2 (1976), https://doi.org/10.1016/0014-4894(76)90127-2

45 Epigenetics studies the many processes and interactions affecting genetic expression. According to Noela Davis, epigenetic research finds

> the patterns of genetic expression that give the distinctive characteristics of a cell or tissue, and thus of the organism, materialise a phenotype in context through a dynamic crosstalk between one molecule and another, body organs, genes and environment, the organism and its surroundings,

in "Material Culture: Epigenetics and the Molecularisation of the Social," in *What If Culture Was Nature All Along?*, ed. Vicki Kirby (Edinburgh: University of Edinburgh Press, 2017).

46 The standard theoretical approach to semiotics is commonly cited as originating with Ferdinand de Saussure's work. For an analysis of Saussurean semiotics that runs counter to the conventional interpretations and aligns Saussure with recent material semiotic thinking, see Vicki Kirby, *Telling Flesh: The Substance of the Corporeal* (London: Taylor & Francis Group, 1997).

47 Increasingly the reach of Saussurean-derived semiotics is being extended to non-humans as studies examine how animals such as dolphins, dogs, corvids, octopi communicate using symbolic languages. Making claims for non-human language use within this tradition has been difficult because of the narrow and demanding criteria for what constitutes a language.

48 Marcel Danesi, "Habit in Peirce: A Review of Consensus on Peirce's Concept of Habit," *Biosemiotics* 11, no. 3 (2018), https://dx.doi.org/10.1007/s12304-018-9334-y

49 Charles Sanders Peirce, The Collected Papers (Historische Texte & Wörterbücher, 1931), Vol V, I, 171, accessed 18 September 2020, www.textlog.de/charles_s_peirce.html

50 Brian Massumi, *Politics of Affect* (Cambridge and Maldon: Polity Press, 2015), 9.
51 C. Barker Jørgensen, "Bivalve Filter Feeding Revisited," *Marine Ecology Progress Series* 142, no. 1 (1996): 228, www.jstor.org/stable/24857244
52 Kohn, *Forests*, chapter 1, Emergent Reals, paragraph 6.
53 Danesi, "Habit", 450.
54 Peirce, The Collected Papers, III, 303.
55 Ibid., III, 324.:

> this notion, of being such as other things make us, is such a prominent part of our life that we conceive other things also to exist by virtue of their reactions against each other. The idea of other, of **not,** becomes a very pivot of thought. To this element I give the name of Secondness.

56 Deacon, "Steps to a Science of Biosemiotics"; Vefa Karatay, Yagmur Denizhan, and Mehmet Ozansoy, "Semiosis as Individuation: Integration of Multiple Orders of Magnitude," *Biosemiotics* 9, no. 3 (2016), https://dx.doi.org/10.1007/s12304-016-9273-4; Kohn, *Forests*.
57 Peirce, The Collected Papers, III, 532.
58 For a concise account of Peirce's understanding of a sign see Mariana Vitti Rodrigues and Claus Emmeche, "Abduction: Can Non-Human Animals Make Discoveries?", *Biosemiotics* 10, no. 2 (2017): 296, https://dx.doi.org/10.1007/s12304-017-9300-0
59 My understanding of these processes was facilitated by Susan Petrilli, *Sign Studies and Semioethics: Communication, Translation and Values* (Boston: De Gruyter, 2014).
60 Karatay, Denizhan, and Ozansoy, "Semiosis", 426.
61 Ibid., 427.
62 Vicki Kirby, "Originary Humanicity: Locating Anthropos," *philoSOPHIA* 8, no. 1 (2018): 55. Kirby does not attribute this model to Peirce, though she also refers to "firstness", so it is possible Peircean logic, and its widespread applications is her consideration here.
63 Kirby, "Originary Humanicity", 57.
64 Vicki Kirby, "Originary Différance: 'A Quantum Vitalism'," *Journal for the Theory of Social Behaviour* 48, no. 2 (2018): 165, https://dx.doi.org/10.1111/jtsb.12164
65 Horn Point Oyster Hatchery, "Oyster Life Cycle," University of Maryland Centre for Environmental Studies, accessed 19 August 2020, http://hatchery.hpl.umces.edu/oysters/oysters-life-cycle/
66 Damian M. Ogburn, Ian White, and Daryl P. McPhee, "The Disappearance of Oyster Reefs from Eastern Australian Estuaries – Impact of Colonial Settlement or Mudworm Invasion?", *Coastal Management* 35, no. 2–3 (2007), https://dx.doi.org/10.1080/08920750601169618
67 Chris L. Gillies et al., "Conservation Status of the Oyster Reef Ecosystem of Southern and Eastern Australia," *Global Ecology and Conservation* 22 (2020): 12, https://dx.doi.org/10.1016/j.gecco.2020.e00988
68 The Nature Conservancy, "Mapping Ocean Wealth: Oyster Reefs," 2019, accessed 6 October 2019, https://oceanwealth.org/ecosystems/oyster-reefs/

69 C. Barker Jørgensen et al., "Ciliary and Mucus-Net Filter Feeding, with Special Reference to Fluid Mechanical Characteristics," *Marine Ecology Progress Series* 15, no. 3 (1984), www.jstor.org/stable/24815945
70 Jørgensen, "Bivalve", 228.
71 Ibid., 289.
72 Jørgensen reports laboratory studies finding that when indigestible and digestible particles were provided to mussels under a range of conditions, there was a 40-80% difference between the numbers of food particles being taken up and the lower numbers of indigestible particles, in Ibid., 290.
73 This language of question and response is in part inspired by Vinciane Despret, "From Secret Agents to Interagency," *History and Theory* 52 (2013): 37–39.
74 Anna E. Murphy et al., "Bioreactivity and Microbiome of Biodeposits from Filter-Feeding Bivalves," *Microbial Ecology* 77, no. 2 (2019): 243–44, https://dx.doi.org/10.1007/s00248-018-01312-4
75 "One oyster is estimated to filter about 180 litres of water per day" according to Jonathan Adams, "Research: Can Restoring Oyster Reefs Combat Nitrogen Pollution?", Cool Green Science, The Nature Conservancy, 2013, accessed 7 May 2020, https://blog.nature.org/science/2013/09/10/restoring-oyster-reefs/
76 Anthony W. Larkum and Ronald J. West, "Long-Term Changes of Seagrass Meadows in Botany Bay, Australia," *Aquatic Botany* 37, no. 1 (1990), https://doi.org/10.1016/0304-3770(90)90064-R
77 Terence Palmer, P. Uehling, and Jennifer Pollack, "Using Oyster Tissue Toxicity as an Indicator of Disturbed Environments," *International Journal of Environmental Science and Technology* 12, no. 6 (2015), https://dx.doi.org/10.1007/s13762-014-0745-2
78 Adams, "Research".
79 Timo Maran, "Semiotization of Matter: A Hybrid Zone between Biosemiotics and Material Ecocriticism," in *Material Ecocriticism*, eds. Serenella Iovino and Serpil Oppermann (Bloomington and Indianapolis: Indiana University Press, 2014), 141.

References

Adams, Jonathan. "Research: Can Restoring Oyster Reefs Combat Nitrogen Pollution?" Cool Green Science. The Nature Conservancy. 2013. Accessed 7 May 2020. https://blog.nature.org/science/2013/09/10/restoring-oyster-reefs/

Barad, Karen. *Meeting the Universe Halfway: Quantum Physics and the Entanglement of Matter and Meaning*. Durham and London: Duke University Press, 2007.

———. "Posthumanist Performativity: Toward an Understanding of How Matter Comes to Matter." *Signs: Journal of Women in Culture and Society* 28, no. 3 (2003): 801–31.

Bennett, Jane. *Vibrant Matter: A Political Ecology of Things*. Durham and London: Duke University Press, 2010.

Bertalanffy, Ludwig von. *General System Theory: Foundations, Development, Applications*. London: Allen Lane The Penguin Press, 1971.

Callon, Michel. "Revisiting Marketization: From Interface-Markets to Market-Agencements." *Consumption Markets & Culture: Agencing Markets* 19, no. 1 (2016): 17–37. https://dx.doi.org/10.1080/10253866.2015.1067002

Chemero, Anthony. "An Outline of a Theory of Affordance." *Ecological Psychology* 15, no. 2 (2013): 181–95.

Coole, Diana and Samantha Frost, eds. *New Materialisms: Ontology, Agency and Politics*. Durham and London: Duke University Press, 2010, Kindle.

Danesi, Marcel. "Habit in Peirce: A Review of Consensus on Peirce's Concept of Habit." *Biosemiotics* 11, no. 3 (2018): 447–52. https://dx.doi.org/10.1007/s12304-018-9334-y

Davis, Noela. "Material Culture: Epigenetics and the Molecularisation of the Social." In *What If Culture Was Nature All Along?* edited by Vicki Kirby, 110–133. Edinburgh: University of Edinburgh Press, 2017.

Day, Timothy Ryan. "With Parted Eye: A Midsummer Night's Dream, Richard Powers' Orfeo, and Biosemiotics." *Green Letters* 22, no. 3 (2018): 247–59. https://dx.doi.org/10.1080/14688417.2018.1512418

Deacon, Terrence W. "Steps to a Science of Biosemiotics." *Green Letters* 19, no. 3 (2015): 293–311. https://dx.doi.org/10.1080/14688417.2015.1072948

Deacon, Terrence W., James W. Haag, and Jay Ogilvy. "The Emergence of Self." In *In Search of Self: Interdisciplinary Perspectives on Personhood*, edited by J. Wentzel van Huyssteen and Erik P. Wiebe, 319–337. Grand Rapids: Wm. B. Eerdmans Publishing Co, 2011.

Despret, Vinciane. "From Secret Agents to Interagency." *History and Theory* 52 (2013): 29–44.

Dolphijn, Rick and Iris van der Tuin. *New Materialism: Interviews & Cartographies*. Ann Arbour: Open Humanities Press, 2012. http://hdl.handle.net/2027/spo.11515701.0001.001

Duke, Norman and Klaus Schmitt. "Mangroves: Unusual Forests at the Seas Edge." In *Tropical Forestry Handbook*, edited by Laslo Pancel and Michael Köhl, 1693–1724. Berlin: Springer-Verlag, 2015.

Gibson, James. *The Ecological Approach to Visual Perception*. Classic ed. New York and London: Taylor and Francis, 2015.

Gillies, Chris L., Sarah A. Castine, Heidi K. Alleway, Christine Crawford, et al. "Conservation Status of the Oyster Reef Ecosystem of Southern and Eastern Australia." *Global Ecology and Conservation* 22 (2020). https://dx.doi.org/10.1016/j.gecco.2020.e00988

Grosz, Elizabeth. *Chaos, Territory, Art: Deleuze and the Framing of the Earth. Wellek Library Lectures in Critical Theory*. New York: Columbia University Press, 2008, Kindle.

Haraway, Donna. *When Species Meet*. Minneapolis: University of Minnesota Press, 2008.

Hoffmeyer, Jesper. "Biosemiotics: Towards a New Synthesis in Biology." *European Journal for Semiotic Studies* 9, no. 2 (1997): 355–76.

———. "Semiotic Scaffolding: A Unitary Principle Gluing Life and Culture Together." *Green Letters* 19, no. 3 (2015): 243–54. https://dx.doi.org/10.1080/14688417.2015.1058175

Horn Point Oyster Hatchery. *Oyster Life Cycle*. University of Maryland Centre for Environmental Studies. Accessed 19 August 2020. http://hatchery.hpl.umces.edu/oysters/oyster-life-cycle/

Hustak, Carla and Natasha Myers. "Involutionary Momentum: Affective Ecologies and the Sciences of Plant/Insect Encounters." *Differences: A Journal of Feminist*

Cultural Studies 25, no. 5 (2012): 74–118. https://dx.doi.org/10.1215/10407391-1892907

International Society for Biosemiotic Studies. What Is Biosemiotics?". Accessed 26 July 2020. www.biosemiotics.org/what-is/

Iovino, Serenella and Serpil Oppermann. "Material Ecocriticism: Materiality, Agency and Models of Narrativity." *Ecozon@* 3, no. 1 (2012): 75–91.

———. "Theorizing Material Ecocriticism: A Diptych." *ISLE: Interdisciplinary Studies in Literature and Environment* 19, no. 3 (2012): 448–75. https://dx.doi.org/10.1093/isle/iss087

Jørgensen, C. Barker. "Bivalve Filter Feeding Revisited." *Marine Ecology Progress Series* 142, no. 1 (1996): 287–302. www.jstor.org/stable/24857244

Jørgensen, C. Barker, Thomas Kiørboe, Flemming Møhlenberg, and H. U. Riisgård. "Ciliary and Mucus-Net Filter Feeding, with Special Reference to Fluid Mechanical Characteristics." *Marine Ecology Progress Series* 15, no. 3 (1984): 283–92. www.jstor.org/stable/24815945

Judith, Kate. "How Mangroves Story: On Being a Filter Feeder." *Swamphen: A Journal of Cultural Ecology* 7 (2020). https://openjournals.library.sydney.edu.au/index.php/Swamphen/article/view/14366

Karatay, Vefa, Yagmur Denizhan, and Mehmet Ozansoy. "Semiosis as Individuation: Integration of Multiple Orders of Magnitude." *Biosemiotics* 9, no. 3 (2016): 417–33. https://dx.doi.org/10.1007/s12304-016-9273-4

Kaygan, Harun. "Material Semiotics of Form Giving: The Case of the Electric Turkish Coffee Pot." *Design Issues* 32, no. 2 (2016): 78–90. https://dx.doi.org/10.1162/DESI_a_00384

Kirby, Vicki. "Autoimmunity: The Political State of Nature." *Parallax* 23, no. 1 (2017): 46–60. https://dx.doi.org/10.1080/13534645.2016.1261661

———. "Originary Différance: 'A Quantum Vitalism'." *Journal for the Theory of Social Behaviour* 48, no. 2 (2018): 162–66. https://dx.doi.org/10.1111/jtsb.12164

———. "Originary Humanicity: Locating Anthropos." *philoSOPHIA* 8, no. 1 (2018): 43–60.

Kohn, Eduardo. *How Forests Think*. Berkeley and Los Angeles: University of California Press, 2013, Kindle.

Larkum, Anthony W. and Ronald J. West. "Long-Term Changes of Seagrass Meadows in Botany Bay, Australia." *Aquatic Botany* 37, no. 1 (1990): 55–70. https://doi.org/10.1016/0304-3770(90)90064-R

Law, John. "Actor Network Theory and Material Semiotics." In *The New Blackwell Companion to Social Theory*, edited by B. S. Turner, 141–158. Oxford: Blackwell, 2009.

Lindeman, Raymond L. "The Trophic-Dynamic Aspect of Ecology." *Ecology* 23 (1942): 399–418. https://dx.doi.org/10.2307/1930126

Maran, Timo. "Semiotization of Matter: A Hybrid Zone between Biosemiotics and Material Ecocriticism." In *Material Ecocriticism*, edited by Serenella Iovino and Serpil Oppermann, 141–154. Bloomington and Indianapolis: Indiana University Press, 2014.

Margulis, Lynn. "Genetic and Evolutionary Consequences of Symbiosis." *Experimental Parasitology* 39, no. 2 (1976): 277–349. https://doi.org/10.1016/0014-4894(76)90127-2

Massumi, Brian. *Politics of Affect*. Cambridge and Maldon: Polity Press, 2015.

Miele, Mara. “The Making of the Brave Sheep or … the Laboratory as the Unlikely Space of Attunement to Animal Emotions.” *GeoHumanities* 2, no. 1 (2016): 58–75. https://dx.doi.org/10.1080/2373566x.2016.1167617

Murphy, Anna E., Rebecca Kolkmeyer, Bongkuen Song, Iris C. Anderson, et al. “Bioreactivity and Microbiome of Biodeposits from Filter-Feeding Bivalves.” *Microbial Ecology* 77, no. 2 (2019): 343–57. https://dx.doi.org/10.1007/s00248-018-01312-4

Odum, Howard T. and Eugene P. Odum. *The Fundamentals of Ecology*. Philadelphia: Saunders, 1953.

Ogburn, Damian M., Ian White, and Daryl P. McPhee. “The Disappearance of Oyster Reefs from Eastern Australian Estuaries – Impact of Colonial Settlement or Mudworm Invasion?” *Coastal Management* 35, no. 2–3 (2007): 271–87. https://dx.doi.org/10.1080/08920750601169618

O’Malley, Maureen A. “The Ecological Virus.” *Studies in History and Philosophy of Biological and Biomedical Sciences* 59 (2018): 71–79. https://dx.doi.org/10.1016/j.shpsc.2016.02.012

Palmer, Terence, P. Uehling, and Jennifer Pollack. “Using Oyster Tissue Toxicity as an Indicator of Disturbed Environments.” *International Journal of Environmental Science and Technology* 12, no. 6 (2015): 2111–16. https://dx.doi.org/10.1007/s13762-014-0745-2

Peirce, Charles Sanders. The Collected Papers: Historische Texte & Wörterbücher, 1931. Accessed 18 September 2020. www.textlog.de/charles_s_peirce.html

Petrilli, Susan. *Sign Studies and Semioethics: Communication, Translation and Values*. Boston: De Gruyter, 2014.

Pickering, John. “Affordances Are Signs.” *Triple C: Cognition, Communication and Co-operation* 5, no. 2 (2007): 64–74. www.triplec.at/index.php/tripleC/article/view/59/61

The Nature Conservancy. “Mapping Ocean Wealth: Oyster Reefs.” 2019. Accessed 6 October 2019. https://oceanwealth.org/ecosystems/oyster-reefs/

van der Tuin, Iris and Rick Dolphijn. “The Transversality of New Materialism.” *Women: A Cultural Review* 21, no. 2 (2010): 153–71. https://dx.doi.org/10.1080/09574042.2010.488377

Vitti-Rodrigues, Mariana and Claus Emmeche. “Abduction: Can Non-Human Animals Make Discoveries?” *Biosemiotics* 10, no. 2 (2017): 295–313. https://dx.doi.org/10.1007/s12304-017-9300-0

von Uexküll, Jacob. “A Stroll through the Worlds of Animals and Men: A Picture Book of Invisible Worlds.” In *Instinctive Behaviour: The Development of a Modern Concept*, edited and translated by Claire H. Schiller, 5–80. New York: International Universities Press, Inc., 1934.

Wheeler, Wendy. “A Connoisseur of Magical Coincidences: Chance, Creativity and Poiesis from a Biosemiotic Perspective.” *Biosemiotics* 7, no. 3 (2014): 389–404. https://dx.doi.org/10.1007/s12304-014-9218-8

———. “Gregory Bateson and Biosemiotics: Transcendence and Animism in the 21st Century.” *Green Letters* 13, no. 1 (2010): 35–54. https://dx.doi.org/10.1080/14688417.2010.10589069

2 Folding and filter-feeding semiotics

As the tide turns, washing across these river mangroves, it folds the river catchment into the Pacific Ocean and the ocean back into the catchment. Strange smells and clouds of summer mosquitoes wash across the urban streets, fish end up on kitchen tables, plastics wash out to be eaten by marine life and pelagic birds or colonised by barnacles, nitrates and eroded shorelines disperse into the ocean. The work of tides folds and intensifies layers of silt; the gravitational attraction to the orbiting moon of earth's water folds waves against the shore, while tectonic forces fold sedimentary rock layers. These earthly folds enact the substance of the coast, folding difference and complexity out of regularity. The folded world is neither firm and inflexible nor fluid and streaming; it is pliable, with ongoing potential for movement and pattern generation. It does not abandon difference but often finds it as a warping rather than a cut and identifies continuity within it.[1]

Folding and involution are approaches to semiotic materiality that avoid the degree of breaching and fixity of differential value that can creep into nested hierarchies of layers. With a fold, there is a direction followed by a turn or reversal of direction. Each meets its neighbour as its inverse – there is an inverse-sameness to the differences of folding. Same meets same and figures the difference between, interfaces emerge from and continue to become through responsive inter-affect.[2] Because of this, there is a semiotic reflexivity inherent to a folded ontology that can be conceived as always remaining within immanence, and twisting or folding it, rather than involving a stepping out from it. The differences that are produced through folding remain relational and sustained by responsive semiotic processes – there is always reading and writing, propositions and responses, continuing across folded interfaces. Folded ontologies often describe a fractal geometry, where folds are folded, again and again, providing an ecology of multiple perspectives.[3] These approaches to material semiotics meld hermeneutics, ontology, epistemology, evolution, and politics together without paradox or aporia. Instead, there is an immanence that twists and folds, so it encounters itself continually. The semiotic material world becomes infinitely reflexively looping with interpretive and abstracting moves. These approaches do not necessarily demolish Peirce's three-part habituation process. Rather, they

DOI: 10.4324/9781003286493-4

remove the sense of a uni-directional progress. This dismantles a hierarchy of levels, rendering the relations of semiotic meaning relative, diverse, and interrelated within a world that is awash with semiosis. In this way, they are more akin to mangrove ecologies.

Folding

Tidal senses

Sydney rock oysters, like many other species of the tidal zone, have developed senses that specifically respond to tidal rhythms.[4] These tidal organisms modulate their physiological processes and behaviour with internal circatidal clocks.[5] The tidal rhythm has become accommodated within the genetic, chemical, and physiological materiality of their bodies so these animals have distinct cellular and physiological mechanisms for their circatidal clocks and for their circadian clocks.[6] All sensing capacities have long evolutionary histories, evolving and shaping through selection, catastrophe, competition, and luck to respond usefully to multifarious inputs. As organisms have increasingly come to be considered as symbiotic assemblages, and as knowledge of the diverse processes through which they gain information has increased, the range of what is regarded as a sense has expanded enormously. Many plant sensitivities have recently been identified.[7] Some mangrove species propagules float upon the tidal currents and do not begin to send down roots until they contact a suitable substrate, sensing those suitable conditions. Sensing can also be dispersed and coordinate without a central control mechanism. This occurs in the sea urchins, whose sensing organs, each functioning in "utter seclusion" from the others, von Uexküll describes as "reflex persons" operating together within a "reflex republic".[8] A mangrove community can, by this reasoning, be described as a body with senses dispersed across the many organisms which participate in its affective capacities.[9] Mangrove ecosystems perceive the tidal rhythm which acts as a coordinating conductor for the mangal reflex republic. A tidal rhythm is sensed and becomes meaningful within the semiotic materiality of the mangal world.[10] The adaptive, responsive development of sense perception has featured as a foundational conversation underpinning many accounts of semiotic materialism. The key insight is a simple one; that perception is always materially and semiotically entangled within the world.

Intertwining, chiasm, flesh

Maurice Merleau-Ponty's investigation of the enfoldedness of perception offers an approach to semiotic materiality that locates meaning within the in-between of inter-responsive perceptual becomings, rather than across abstracted layers of discernment performed by habituation. This approach does not operate as a substitute to a Peircean model, because it is not a semiotic theory in the same sense. Rather, it has been used to complicate and complement semiotic models by suggesting an increased focus on the ongoing processes at play throughout material semiosis, troubling any distinctions into separate parts or stages. The sensing body, as Merleau-Ponty describes it, is always in-between subject and object, performing a continual shifting between. This strange double focus is most famously explored through a discussion of the relationship that occurs when his one hand touches his other hand, which touches an object such as the tabletop.[11] In this move, he experiences being both the object which is touched and the subject who touches. The experience is both one and two, and subject and object are intertwined within the experience/s. The two roles do not dissolve into one. Indeed, they remain strangely distinguishable, yet neither is the experience of each detachable from the other. They occur together, taking their forms as separate experiences in continuous jittery response to each other. There is a particularity to this interconnectivity that is also a gap; it/they is/are lively with co-becoming, continually attending to each other, reading their forms from responsiveness to each other. This is why Merleau-Ponty employs the terms "flesh", "intertwining", and "chiasm", to move away from language which hearkens back to modes of thinking that demand detachment between what is sensed and the sensing of it. Here there is neither two nor one, but a jittery intertwining or criss-crossing, neither separation nor congruity but the uncertainty of chiasm: a doubled and twisting fold.

Body and world are conjoined as "flesh" in Merleau-Ponty's conceptualisation. Flesh is the whole within which sensing, perceiving beings are enfolded chiasmatically with that which affects and shapes them, which they in turn affect. Grosz describes flesh as "the shimmering of the world's difference from itself and thus the subject's difference from itself".[12] In the example of the two hands, flesh is touching as well as tangible, the two modes of being bringing each other into meaningfulness as they become distinguishable. Merleau-Ponty picks out the reversibility of these interacting modes as the essence of flesh, which is "one sole movement in its two phases".[13] The concept of flesh grasps the thickness of this gap; "all the thickness of the body" performs the meaningful world,[14] what is between cannot be viewed as a clear step between levels because peering into the process at any one of these steps, there is always this strange circularity of the one that is two.

This reversibility plays a key role as Merleau-Ponty derives a narrative account for an inherent semiotic meaningfulness from the intertwining sensations of the flesh of the world. As he finds his body as something both tangible and touching, the two movements "incorporate themselves into the universe they interrogate, are recorded on the same map as it; the two systems are applied upon one another, as the two halves of an orange".[15] Each needs the other and is made meaningful through the intertwining. He is tangible and touching within a universe wherein these phenomena map each other into significance; a universe that comes into meaningful existence through that criss-crossing, each movement providing a question into which the other can come to matter as response. How edges, beings, shapes, durations, diffraction patterns, are shaped is a continuing process of experimental figuring among the materiality of the world and the diverse sensing apparatus the world itself produces (that is to say; within flesh): "self and content are intertwined, each providing a place for the other to appear".[16] Beyond just affecting the sensitivity of a subject's capacity to know, this kind of looping reflexivity is a maker of meaningful worlds.

Although intertwined, the sensing body and the sensed world are not one same thing. The mismatch between them is generative. Perception occurs across and creates difference.[17] This gap is thick with the flesh of the world: "the thickness of flesh between the seer and the thing is constitutive for the thing of its visibility as for the seer of his corporeity; it is not an obstacle between them, it is their means of communication".[18] Later Merleau-Ponty writes: "this hiatus [...] is the zero of pressure between two solids that makes them adhere to one another".[19] It is nothing at all and yet also the source of the differences which form and shape things. This gap, thick as the world and thin as the zero pressure of complete adherence, is the interface across which a meaningful world and an interpretive perspective become in response with each other. It is a foldedness or twisting that brings about a discursively inter-responsive world.

As the wave moves across the mud, in that ebb and flow, the touch between sea and land shape each other. The sea held and moved into its form by the touch of the land, the land built, eroded, and shifted through the touch of the sea. The mangrove thickens in the chiasm of this touch. This chiasm is neither smooth nor simple. It is a relation of affective friction that rubs and wears, and that restrains and allows, the flow and the edge co-becoming, shaping each other. Where there is flow, there is constraint and also the release from it. Flow is as it is by its constraints, constraint is as it is by the action of the flow through it. It is only flow insofar as it moves without boundaries, yet only flow as it is bound. There is a chiasm folded within every flow.

Merleau-Ponty describes the flesh of the world as folded in statements such as: "the flesh we are speaking of is not matter. It is the coiling over of the visible upon the seeing body, of the tangible upon the touching body".[20] For Merleau-Ponty folding is the mechanism and the inherent quality of a semiotic materiality, one that reads and writes itself, fold upon fold, meaning and matter, "irreducibly interwoven" and "immanently generative".[21] A folded ontology twists together spatiality and temporality; a twisting through which becoming and the perspective from which to witness it both occur. Responses fold upon each other again and again through the mangrove, echoing through biology and geology, becoming intricate, diverse, and complex.

Folds in Deleuze, Haraway and Barad

In *The Fold*, Gilles Deleuze explores the Baroque's interest in the world as folded, particularly as this is revealed within Leibniz's cosmology. The

Baroque "twists and turns the folds, takes them to infinity, fold upon fold", becoming fascinated by the artistic and intellectual possibilities of a folded materiality and metaphysics.[22] Leibniz, in Deleuze's analysis, envisages a dualistic world of two parts, infinitely and complexly inter-folded in this baroque manner. One part is interiority and immanence. This is the infinitely-folded monad, the world without absolute discontinuities. Deleuze associates it with an idealised Baroque interior, illuminated only by mirrors, joyously and eternally self-referential. The other part is an interpretive frame that discerns physical laws and describable phenomena. It is the site of exteriority that only cares to describe facades, which can then go on to circulate as signifiers. These two are "infinite sets which never meet" and yet they are entirely inter-folded.[23] Folding permits this kind of separation-in-association; the sort of relationship developed in Merleau-Ponty's intertwining and chiasm and tidally embodied as the chiasmic relations between currents, oysters, and mangal mud.

Deleuze develops a view of an "ontological fold" that maintains much of his interpretation of the Leibnizian model but, additionally, presents the interpretative, folding interface as critical and evaluative.[24] Phenomenological exteriority, the experience of the existence of bounded entities or concepts, he proposes, should be considered as problems or questions. Taking in evidence and constructing a facade to account for it is a way of asking; "why/how is there this thing, concept or event?", and "what is this thing or concept or event?". This questioning perspective folds back into the ongoing immanent-being-as-folding-processes of the world, "as though there were an 'opening', 'a gap', an ontological 'fold' which relates being and the question to one another".[25] The discernment of difference and its immanent becoming within the world, fold and unfold against each other, each becoming through a question and response each to the other,[26] the facade as a questioning and immanent becoming meeting it in response.

Within this dynamic folding, the self occupies the vanishing interface, where point of view becomes in inter-responsiveness with that which it brings to focus. John Wylie explains that for Deleuze: "The self [...] is that apprehensive ability which emerges to occupy a point of view established through the variations and inflections of folding".[27] Self is not a fixed perspective, but an ongoing relationship across a fold in the world, a fold that effects what Wylie describes as the landscape (we could call it the discerned, or the facade) and the self that distinguishes it. "Relations between selves and landscapes are motile relations, an incessant movement of enfolding and unfolding, openness and enclosing, in which the two implicate (fold with) and include each other".[28] Perceived and perceiver are enfolded in the other, becoming together. A folded ontology twists together spatiality and temporality; a twisting through which becoming and the perspective from which to witness it both occur. The self is oyster-like, bringing the world to bear as a fold in it, and both surrounded and inhabited by other selves.

Other writers seeking ways to explain immanent differentiation have taken up this visualisation of immanent becoming as enfolding affect.

Donna Haraway, working within the parameters of biology, refers to the enfoldedness of symbiosis as the underlying process through which life becomes more complex. Life forms enfold each other, becoming obliged within symbiotic relationships; "irresistible attraction toward enfolding each other is the vital motor of living and dying on earth."[29] Karen Barad elaborates a semiotic material ontology beyond the biological. She writes, for example, of the universe as the enfolded memories of matter: "Memory – the pattern of sedimented enfoldings of iterative intra-activity – is written into the fabric of the world. The world 'holds' the memory of all traces; or rather, the world is its memory (en-folded materialisation)".[30] Barad attempts, as Deleuze did, to stay within immanence, to avoid a split in the world, writing of intra-activity, intra-relatedness, intra-agency: "Becoming is not an unfolding in time, but the inexhaustible dynamism of the enfolding of mattering".[31]

From Merleau-Ponty to Deleuze and on to Barad, the fold is invoked to describe the relationship between immanence and emergence across the boundary of the event of discernment. By immanence, I have in mind intra-affect, always in process, on the move and unaware of defined boundary limitations or goals. By emergence, I mean the coming into definition, from immanence, of bounded concepts – bodies, species, assemblages, methods, temporalities, and so on – that are discerned by something. This discernment can take many forms and may or may not involve cognition. It may take the form of anticipation of tidal patterns based upon past experiences, materialised in the pneumatophores of a mangrove tree and the muscular movements of an oyster; or it may involve a new material affect requiring a response, as the mangrove and oyster cells respond to the presence of petrochemicals and plastics. Folding is a middle way that combines structure and process and does not deny layering but emphasises that any layers are emergent within the doing of folding. Here, as patterns such as edges, bodies, habituated processes, assemblages, and narratives of relations emerge, so do the criteria supporting their meaningfulness.

Across this fold, within tidal zones, a mangal is the negotiations between the immanence and the emergence of its becomings. That is, it is becoming through the interactions between, on the one hand, what comes bubbling out from between interactions amongst many inter-affective trajectories (immanently), and the many ways "mangrove" is thought or experienced to be, by whatever has a sense of it, on the other (emergently). The quiveringly political nature of the thick gap of the in-between is apparent within this way of knowing.

Further concepts of the fold

Involution

Several concepts from the literature have developed a folded material semiotics in useful directions for a post-anthropocentric project. Involution is a variation on folding with etymological reminders of its evolutionary, collaborative, and embodied qualities. Deleuze and Guattari invoke the term as an alternative to "evolution" that emphasises the creativity and play of experimental becoming, and specifically deemphasises the delimitations of levels.[32] Carla Hustak and Natasha Meyers explore the concept further, pushing into the processes occurring as beings become involved together within the interstices of their relations. They describe involution as a momentum, like that of dancers, whose movements are at once anticipatory, responsive, and directive, operating in an "affective push and pull among bodies, including the affinities, ruptures, enmeshments, and repulsions among organisms".[33] Hustak and Meyers follow Darwin and Deleuze and Guattari and apply their involutionary reading to the *Ophrys* orchid and wasp[34] relationship of entanglement. Their approach aims to stay with the relationship as it is in play "'now' and 'now' and 'now'",[35] finding it improvised and experimental and that emergent entities are not stable, but of the moment, fleeting, and contingent. "Instead of two autonomous, clearly demarcated individual organisms engaging in a sexual encounter", they write, "the wasp-orchid is an 'intra-active' phenomenon".[36] All the interest, all there is, is in the communicative between. Material perception, interpretation, and translation – semiotic materialism as the attunement of sensitive and sensory bodies – fill the entire space.

Vinciane Despret also develops a theory of involution, expanding on Isabelle Stengers' usage. Stengers and Despret acknowledge a somewhat firmer maintenance of identities than Deleuze and Guattari and Hustak and Meyers, emphasising that these identities are "transformed but not confused: each critter still differentiates, but differentiates differently".[37] With Michel Meuret, Despret explores involution through the becomings of flocks of sheep, inclusive of their shepherds, sheepdogs, wolves, and a range of plants. The flock comes into its particular space and time as a composition,[38] crafted by the actions of each individual but only in so far as these are always interactions, as each participant attends to the others and makes accommodations and suggestions, accepts or rejects or amends proposals, forms expectations based upon past occurrences, and eventually comes to accede to flock norms. While the flock has an emergent presence, in that it comes to have its own space and time, this is never a "black box", but always a delicate practice of inter-responsive composition which also involves the meadow vegetation, the topography of the hillsides and waterways, and the varied temporalities of the landscape. There is an aesthetic in play as the shepherds develop a sense of what it means for their flocks to be "beautiful". A flock becomes more beautiful as it develops norms that are increasingly respectful of the feelings of each member; a flock is beautiful to the extent that the attentions of each member are towards the suggestions and responses of the others, and towards the flock's traditions. The emergent entity, the flock, is a delicately performed composition occurring only because of the sustaining of ongoing immanent processes that reflect those respectful norms. Such sustainability is effortful, like the dancer who sustains a pose only through the continuing strain of her muscles.[39] Emergence occurs by virtue of this ongoing intra-attentiveness and, where there is beauty to be found, it is emergent from a combination of tenderness and aptness among immanent processes.

Despret also explores the example of the orchid and the bee, where very different beings are "getting-involved" in making themselves together within responsive relations.[40] She adds into that involution the researchers who have historically investigated this relationship and finds no reason why that involvement should end at the laboratory door. Darwin and all the other biologists, socio-biologists, and ethologists who have studied these orchid-bee relationships have also been involving themselves, changing themselves in response to these organisms, and affecting the relations through their work. She writes: "Each living being renders other creatures capable (of affecting and of being affected), and they are entangled in a myriad of rapports of forces, all which are 'agencements' ".[41] Some beings are "capable of making other beings capable", and in doing so, they propagate agencement, the force that invokes agency.[42] Here agency is something beings gift to each other through sufficiently attentive involution. As such, it is both fragile and borrowed, while also becoming strengthened as beings attend to each other in ways that enhance capacities. Despret notes that this means agency can be lost when thick inter-relationalities are lost. She argues that plants long cultivated by gardeners lose the ability to propagate themselves because they are no longer able to "pay attention" to the changing ways of the other species and individuals around them: "They have been rendered 'deaf' and 'mute' through intensive agricultural practices and pesticide use".[43] The same claim can be made for animal breeds who have lost the ability to function except within the care of humans – extreme toy dog breeds or fancy fish and fowl breeds, for example – and perhaps of the growing number of species maintained predominantly through captive breeding programs because of the loss of their habitat and their near extinction.[44] This is a vision of a semiotic materiality that is strengthened and made more beautiful by the growing thickness of the complexities of networks of inter-attentiveness but equally diminished and made barren as these complexities thin.

Synecdoche and refrain

The criss-crossed, intertwining, reversibility of flesh demonstrated here resonate with the concept of ecological synecdoche explored by Michael Marder[45] and Delia Nasser and Margart Barbour.[46] A synecdoche occurs within language when some part of a concept expresses the whole entity within it. A tree, Nasser and Barbour suggest, is a two-way, reversible synecdoche of its environment because its body is made of the stuff of that place and takes the form it does in response to the particular material conditions of its location. The environment, in turn, expresses the tree, which provides the humus and mulch and other non-mineral physical structures (trunk, branches, roots) that comprise it; "the very notion of 'environment' depends on and cannot be separated from those who actively participate" in it. "Trees express their environments in their form and activity; and the environment is expressed (realised) in and through the trees". The two "emerge simultaneously and in relation to one another". A tree "emerges as a whole only in its environment", its structure is "dialogical, responsive and fluid".[47] Oyster and mudbank likewise can be described as co-expressed within an ecological synecdoche wherein each writes the other. Although Nasser and Barbour do not use the term, their discussion sits very comfortably within a semiotic material ontology.

Deleuze and Guattari's work on the refrain recognises rhythm as ontologically foundational. The refrain is a way of thinking about how the possibilities that rhythm proposes are replied to in ways that carve out territories of meaningfulness.[48] It is the iteration of a rhythm into some kind of motif that answers back to the question that the rhythm proposes; "the becoming-expressive of rhythm, the emergence of expressive proper qualities, the formation of matters of expression that develop into motifs and counterpoints".[49] The crab, the mud whelk, and the mangrove are material refrains of the tidal rhythm, the creative work and play of that rhythm of the world upon itself, in answer to itself, gradually, experimentally, and artistically becoming differentially articulated into diverse forms. These beings have territorialised tidal rhythm into their bodies, their histories, their behaviours, and their relations.

Because they interrelate and affect each other, organisms are part of many refrains, those that encompass each relationship, and those that encompass multiple relationships. Refrains are occurring among genetic rhythms at a species level, and across multiple individuals and species within ecological communities. "The tick and the mammal whose blood it extracts, the spider and the fly it captures, are contrapuntal or harmonic forces, duetting features that must be considered as part of one and the same refrain", Grosz writes, in examples of refrains within particular relationships shaped by evolutionary processes.[50] Refrains are inter-affecting, of the between. The mangroves at Lime Kiln Bay bring together the weavings of sharks, the drifting of mud, the search for shelter of fish, the hunger of cormorant, the cellular urge to grow of a mangrove seedling, each with its own rhythms, which interweave with the rhythms of many others, all paced together with tidal and diurnal rhythms, rhythmically entangled as mangrove. "Fundamentally constructive", a refrain brings together "disparate elements all fundamentally vibrational [...] into an organized synthetic totality".[51] It is "intra-assemblage",[52] the responsive glue holding together various bodies, temporalities, spatialities, inter-responsively aligning them as refrains of each other, into the mangrove assemblage.

In becoming entangled new possibilities for ways of becoming are created, and others are lost, changed, or curtailed. That is why Deleuze and

Guattari believe a refrain includes processes of deterritorialisation within it, to disrupt and keep lively. If a refrain takes shape in response to regularity, (the processes of territorialisation), that shape is continually required to refine itself in response to disturbances of that regularity (the process of deterritorialisation). The oyster is, by this account, a refrain continually territorialised by the regularity of the tide and deterritorialised by its irregularity. The deterritorialising is like a motivator of change, the territorialising the attractor towards stability. A refrain restricts as well as enables, it makes demands and obliges bodies into complex entangled interdependent stories.

Affect

Affect is another concept that has been important within the grammar of material semiotics which invites an understanding of material semiotics as inter- or intra- responsive and chiasmatic.[53] Here I am thinking of the conceptualisation of affect as employed by Bergson, Whitehead, Deleuze and Guattari, and particularly Massumi, that might be described as affective materiality. Barad describes a very similar concept as "intra-action". Affect here is understood as both "causing a difference (to something)" and "being changed (by something)". The "somethings" are bracketed because their existence is not relevant to the definition; indeed, they may not exist prior to an affective event and, if they do, they will be changed by it. It is more accurate to also bracket out the "causing" and simply equate affect with differencing as an action, which emphasises it as intra-affect. Affective materiality is thus the idea that the material world is riven by affect – that one process of differing or change sets the next in motion so "something new" continues to be possible.[54] The mudbank at the river's edge, the tidal current, the oyster, and the river flow, all affect each other, and in so doing so all become, and continue becoming, together.

The same term, affect, is often used to describe both sides of the relationship, the affecting and the being affected, a usage which Massumi attributes to Spinoza's definition of affect as "the power to affect and be affected".[55] As Massumi explains, Spinoza's definition captures an elegant and appropriate sense of reciprocity. It makes affect a quality of openness to relationship, rather than being about change within an already constituted thing. Thinking of affect this way emphasises tensions across this ontological gap, waverings in the face of the power of affect. It suggests an uncertainty about where and when and how the power to affect or be affected will become an action and infuses the between with tensions about potential power. Affect imbues the world with the lurking tension of potential. Massumi thus calls affect "proto-political" and finds within it the germ of a politics that infuses the living world.[56]

Where there is affect, there is mattering in Barad's doubled/material semiotic sense of having a performative material impact.[57] Difference, as the basis for the patterning-up of the world, for the coming into being of the distinctiveness between things, emerges from affect. Difference certainly matters, it begins and ends things, or suggests their beginnings and ends, but difference itself emerges from process and includes uncertainty and proto-political power.[58] Things do not begin and end neatly; there is interplay, tension, negotiation, figuring-out, adaption, argument, and at times impasse, within affect. Because of the redoubled nature of affect – each affect also being prehension to the other's affect – this relationship is lively, in intra-adaptive chiasmic tension. This is why semiotic materialism finds that the patternings and relations of materiality become from within the process of mattering itself, as the mudbank and the oysters come through their interrelations within filter-feeding.

Affect, ecological synecdoche, refrain, flesh, filter-feeding, chiasm, mangaling: this discussion is collecting a lexicon and a conceptual grammar for discussing reflexively folded interstitiality understood within a non-anthropocentric material semiotics.

Oyster semiotics

Oyster semiotics

What happens if we think of semiotics more like filter-feeding as it is performed through the thickness of oyster bodies in the intertidal? The filter-feeder depends upon, indeed is fundamentally entangled with, the unfiltered potentiality of the tidal flow and cannot continue without eliminating what its filtering body determines as waste materials. It becomes responsively into the betweenness of a filter as it builds, through its bodily processes, the categories that constitute that between. This is the case both with respect to its materiality as the highly specialised species form and behaviours it inherits, as well as into its momentary practices of living. The oyster becomes complex and expert through ongoing experimental response, filtering on the move, or along an unfolding trajectory or line.[59] This is not a static filter with fixed criteria of sorting. The criteria are figured responsively and adaptively within the process, as the body itself. This requires a filter-body that remains vulnerable in order to figure how, in the experimental abductions of the moment, to respond sufficiently to survive, perhaps to exploit, or to transform. The parameters for the development of criteria are also in negotiation, emergently becoming within the processes. Such a vulnerability is never innocent, however, because it is generated within situated experience.[60] As emergent of experience, it is the opposite of innocent. That non-innocent vulnerability makes for a filter that may respond creatively, bringing to bear the experience as well as the immediacy of need. Creative evolution and responsive interaction produce tiny cirri that make complex coordinated movements, mucus nets, finely curving mantles, lucent pearls, microbe-rich blobs of pseudofaeces, and the geobiome of the muddy shore.

Considered as filter-feeder, a subject is a creative and adaptive organiser and sorter of worlds/narrative meanings who is entirely inter-responsive with the world that it shapes. Any intercession changes subject and object together. Introduce toxins or plastics and the oysters change materially, as will the composition of the mud the oyster builds through its filtering. The Sydney rock oyster, for example, develops an often-fatal shell deformity when exposed to the chemical tributyltin used in boating[61] and many chemical pollutants affect oyster reproduction.[62] If they can manage to reproduce, oyster offspring may be differently composed,[63] potentially altering the population into the future. The filter-feeder renders difference into its world and is fundamentally shaped as a dynamic responsive subject by that rendering.

Filtering and filter-feeding

Filter-feeding is a transformational process. Debris swept on the tidal flow is transformed, through complex biological processes that gather, select, digest, and expel, into the diverse cell types, organs, and physiological processes of oyster bodies and into the excreted undigested material, shaping bodies and environments. By contrast, filtering as it is practiced within many human contexts seems reductive rather than transformative. In a chemistry laboratory, for example, filtering is a selection process based upon criteria set by the scientist. Unlike the evolving and responsive interactions between the body of the oyster and the flow of the particles, the filtering here seems to be constrained within the parameters of a predetermined question asked by only one of the participants.[64] The criteria for the differentiation are selected for by the scientist, who chooses the filter. However, as science and technology studies scholars like Latour, Despret, and Haraway have shown, the apparatus of science are not passive.[65] They speak back to the actions of the scientists; their signatures are deeply imprinted upon the questions it is possible to ask, and the understandings that emerge. In both the mangrove and the laboratory, there is something happening to the criteria of enquiry across the filtering process that impacts upon all the participants. In the oyster, cirri twist in motions that slow down and speed up the fluid flow in particular ways, mucus creeps down towards the exit point for the pseudofaeces. In the laboratory, scientist and filtering apparatus coordinate and collude to select the desired filtrate and discard the undesired. Perhaps filter-feeding is a more appropriate metaphor for research in general, showing its thickness, creativity and folded reflexivity.

Notes

1 Diana Coole describes this intra-responsive folded world: "As matter swirls and metamorphoses, its modulations do yield provisional forms as styles of existing. But these also subdivide into increasingly tiny folds that sustain their internal

integrity across a continuous fabric of folds within folds", in "The Inertia of Matter and the Generativity of Flesh," in *New Materialisms: Ontology, Agency, and Politics*, ed. Diana Coole and Samantha Frost (Durham and London: Duke University Press, 2010), 113.

2 For example, Coole writes that folds envisage "collective life as a complex relational field that emerges in an intercorporeal, intersubjective 'between'", "Inertia".

3 For example: Gilles Deleuze and Felix Guattari, *A Thousand Plateaus: Capitalism and Schizophrenia*, trans. Brian Massumi (Minneapolis and London: University of Minnesota Press, 1987), 286.

4 Organisms with internal circatidal clocks include many crustacean species, molluscs, fish, and some insects such as the Asian mangrove cricket and some mangrove-dwelling midges.

5 Amorina Kingdom, "Creatures with Circatidal Rhythms Move to a Different Beat", *Hakai Magazine*, 2020, www.hakaimagazine.com/news/creatures-with-circatidal-rhythms-move-to-a-different-beat/; David Wilcockson and Lin Zhang, "Circatidal Clocks," *Current Biology* 18, no. 17 (2008), https://dx.doi.org/10.1016/j.cub.2008.06.041

6 Wilcockson and Zhang, "Circatidal Clocks".

7 Accounts of plant senses include Michael Marder, "Plant Intentionality and the Phenomenological Framework of Plant Intelligence," *Plant Signaling and Behavior* 7, no. 11 (Nov 2012), https://dx.doi.org/10.4161/psb.21954; Carla Hustak and Natasha Myers, "Involutionary Momentum: Affective Ecologies and the Sciences of Plant/Insect Encounters," *Differences: A Journal of Femininst Cultural Studies* 25, no. 5 (2012): 74–118; Tony Trewavas, "Plant Intelligence: An Overview," *BioScience* 66, no. 7 (2016), https://dx.doi.org/10.1093/biosci/biw048

8 Jacob von Uexküll, "A Stroll through the Worlds of Animals and Men: A Picture Book of Invisible Worlds," in *Instinctive Behaviour: The Development of a Modern Concept*, ed. and trans. Claire H. Schiller (New York: International Universities Press, Inc., 1934), 5–80.

9 According to the ecologist Maureen O'Malley:

> ecology relies on a distributed notion of individual, in which biological coherence and activity are distributed across collections of single organisms and their parts. These distributed agents have causal effects, but that causality is the aggregated activity of a loosely bounded collectivity: the distributed individual,

in "The Ecological Virus," *Studies in History and Philosophy of Biological and Biomedical Sciences* 59 (2018): 77.

10 Kate Judith, "How Mangroves Story: On Being a Filter Feeder," *Swamphen: A Journal of Cultural Ecology* 7 (2020). In this article, I discuss mangrove dispersed perception.

11 Maurice Merleau-Ponty, *The Visible and the Invisible* (Evanston: Northwest University Press, 1968), 9, 133–41, 48–50. This example of the two hands touching is also important in Merleau-Ponty's earlier text, *The Phenomenology of Perception*, trans. Donald A. Landes (Oxon and New York: Routledge, 2012), Kindle. Merleau-Ponty inherited the example from Husserl.

12 Elizabeth Grosz, *Time Travels: Feminism, Nature, Power* (Crows Nest: Allen & Unwin, 2005), 125.
13 Merleau-Ponty, *Visible*, 138.
14 Ibid., 136.
15 Ibid., 133.
16 James Mensch, "Arousal: The Intertwining of the Within and the Without," *World Picture* 4 (2010): 6, www.worldpicturejournal.com/WP_4/Mensch.html
17 Merleau-Ponty, *Visible*, 147.
18 Ibid., 135.
19 Ibid., 148.
20 Ibid., 146.
21 Coole, "Inertia", 101, 04.
22 Gilles Deleuze and Jonathan Strauss, "The Fold," *Yale French Studies* 80 (1991): 277.
23 Deleuze and Strauss, "The Fold", 233.
24 Gilles Deleuze, *Difference and Repetition*, trans. Paul Patton (New York: Columbia University Press, 1994), 76.
25 Deleuze, *Difference*.
26 Ibid., 236.
27 John Wylie, "Depths and Folds: On Landscape and the Gazing Subject," *Environment and Planning: Society and Space* 24, no. 4 (2006): 531, https://dx.doi.org/10.1068/d380t
28 Wylie, "Depths and Folds".
29 Donna Haraway, "Staying with the Trouble for Multispecies Environmental Justice," *Dialogues in Human Geography* 8, no. 1 (2018): 58.
30 Karen Barad, "Nature's Queer Performativity", *Qui Parle: Critical Humanities and Social Sciences* 19, no. 2 (2011): 145–46. See also Karen Barad, *Meeting the Universe Halfway: Quantum Physics and the Entanglement of Matter and Meaning* (Durham and London: Duke University Press, 2007), 8.
31 Barad, *Meeting*, 609.
32 Deleuze and Guattari, *Plateaus*, 164, 238, 70.
33 Hustak and Myers, "Involutionary", 97.
34 Many species of the genus *Orphrys*, commonly known as Bee Orchids, produce flowers that are physiologically shaped to sexually attract bee pollinators. Some release chemicals that inspire sexual attraction, and many provide flowers that stimulate the bees mating behaviour. During mating, the flowers deposit pollen on the bees or wasps, for distribution to the next flower. Used as an example of evolutionary co-adaption by Darwin, and revisited by Haraway and others, the bee-orchid and the bee or wasp have become the "go-to" example of evolutionary attunement.
35 Hustak and Myers, "Involutionary", 97.
36 Ibid.
37 Vinciane Despret and Michel Meuret, "Cosmoecological Sheep and the Arts of Living on a Damaged Planet," *Environmental Humanities* 1, no. 8 (2016): 31–32, https://dx.doi.org/10.1215/22011919-3527704
38 Despret and Meuret, "Cosmoecological Sheep", 32.
39 Helena Bender and Kate Judith developed an approach to sustainability based upon its emergence from mutually nourishing immanent inter-responsive relations, in "Does Sustainability Emerge from between the Scales?",

Emergence: Complexity & Organization 17, no. 1 (2015), https://dx.doi.org/10.emerg/10.17357.69fc4cb80d37cf839b28994be1b20259

40 Despret is highly critical of the tradition among socio-biologists to scour these relationships of any subtlety and activity involved in the mutual ongoing attunement between partners and view them only in terms of calculated trickery, as strategically concerned with exploitation and deception. See Vinciane Despret, "From Secret Agents to Interagency," *History and Theory* 52 (2013): 29–44.

41 See Despret, "Secret", 37.

42 Ibid., 38.

43 Ibid., 37.

44 See, for example, Thom van Dooren, *Flight Ways: Life and Loss at the Edge of Extinction* (New York: Columbia University Press, 2014), chapter 4, Kindle. Here van Dooren explores the limited opportunities available in the lives of cranes in a captive breeding program.

45 Michael Marder, *Plant-Thinking: A Philosophy of Vegetal Life* (New York: Columbia University Press, 2013).

46 Delia Nassar and Margart Barbour, "Rooted", *Aeon Ideas*, 16 October 2019, https://aeon.co/essays/what-can-an-embodied-history-of-trees-teach-us-about-life

47 Nassar and Barbour, "Rooted".

48 Deleuze and Guattari, *Plateaus*, 313.

49 Ibid., 322.

50 Elizabeth Grosz, *Chaos, Territory, Art: Deleuze and the Framing of the Earth.* (New York: Columbia University Press, 2008), 53.

51 Grosz, *Chaos*, 56.

52 Deleuze and Guattari, *Plateaus*, 323.

53 Considering material semiotic becoming as a two-in-one folding or criss-crossing within/between immanence and emergence, these two, inter-response and intra-response, are no longer distinct – in this context, each implies the other and they are neither-one-nor-two because they occur and become together across the affective event.

54 According to Brian Massumi:

> From the point of view of a given context, affect is the quasi-causal openness of a characteristic interaction under way in that context to a sensing of "something new," the arrival or irruption of which is expressed in a global qualitative change in the dynamic of the interaction,

in *Parables of the Virtual: Movement, Affect, Sensation* (Durham and London: Duke University Press, 2002), 227.

55 Brian Massumi, *Politics of Affect* (Cambridge and Maldon: Polity Press, 2015), 48.

56 Massumi, *Politics of Affect*, 3.

57 Barad, *Meeting*, 182.

58 Deleuze, *Difference*.

59 See Tim Ingold. *The Life of Lines* (Oxon and New York: Routledge, 2015), 16.

60 Donna Haraway frequently emphasises this non-innocence of vulnerability when regarded within relational ontologies where more attentive responsiveness makes for more apt responses. See, for example: "The Promises of Monsters: A Regenerative Politics for Inappropriate/D Others," In *Cultural*

Studies, ed. Lawrence Grossberg, Cary Nelson, and Paula A. Treichler (New York: Routledge, 1992), 297; Haraway, "Trouble", 29; Haraway, *When Species Meet* (Minneapolis: University of Minnesota Press, 2008), 128.

61 Gavin. F. Birch, Marcus. S. Scammell, and Colin. H. Besley, "The Recovery of Oyster (Saccostrea Glomerata) Populations in Sydney Estuary (Australia)," *Environmental Science Pollution Research* 21, no. 1 (Jan 2014), https://dx.doi.org/10.1007/s11356-013-2168-x

62 Katelyn Edge et al., "Indicators of Environmental Stress: Cellular Biomarkers and Reproductive Responses in the Sydney Rock Oyster (Saccostrea Glomerata)," *Ecotoxicology* 21, no. 5 (Jul 2012): 1416, https://dx.doi.org/10.1007/s10646-012-0895-2

63 Edge et al., "Indicators of Environmental Stress", 1422.

64 Vicki Kirby makes this point in relation to scientific practice in general:

> Whereas authorship in nature could be regarded as dispersed and local at the same time, the tripartite parsing of inquiring subject, object, and mediating apparatus, precludes such a possibility. In this scene of scientific inquiry, the one who initiates the question, by definition, cannot be the object or the apparatus,

in, "Originary Humanicity: Locating Anthropos," *philoSOPHIA* 8, no. 1 (2018): 55.

65 The point is well developed in Donna Haraway, *Modest_Witness@Second_Millennium. Femaleman_Meets_Oncomouse: Feminism and Technoscience.* (New York: Routledge, 1997).

References

Barad, Karen. *Meeting the Universe Halfway: Quantum Physics and the Entanglement of Matter and Meaning*. Durham and London: Duke University Press, 2007.

———. "Nature's Queer Performativity." *Qui Parle: Critical Humanities and Social Sciences* 19, no. 2 (2011): 121–58.

Bender, Helena and Kate Judith. "Does Sustainability Emerge from between the Scales?" *Emergence: Complexity & Organization* 17, no. 1 (2015). https://dx.doi.org/10.emerg/10.17357.69fc4cb80d37cf839b28994be1b20259

Birch, Gavin. F., Marcus. S. Scammell, and Colin. H. Besley. "The Recovery of Oyster (Saccostrea Glomerata) Populations in Sydney Estuary (Australia)." *Environmental Science Pollution Research* 21, no. 1 (Jan 2014): 766–73. https://dx.doi.org/10.1007/s11356-013-2168-x

Coole, Diana. "The Inertia of Matter and the Generativity of Flesh." In *New Materialisms: Ontology, Agency, and Politics*, edited by Diana Coole and Samantha Frost, 766–773. Durham and London: Duke University Press, 2010.

Deleuze, Gilles. *Difference and Repetition*. Translated by Paul Patton. New York: Columbia University Press, 1994. Originally published as *Différence et Répétition*, Presses Universitaires de France, 1968.

Deleuze, Gilles and Felix Guattari. *A Thousand Plateaus: Capitalism and Schizophrenia*. Translated by Brian Massumi. Minneapolis and London: University of Minnesota Press, 1987. Originally published as *Mille Plateaux: Capitalism et Schizophréne 2*, Les Éditions de Minuit, 1980.

Deleuze, Gilles and Jonathan Strauss. "The Fold." *Yale French Studies* 80 (1991): 227–47.

Despret, Vinciane. "From Secret Agents to Interagency." *History and Theory* 52 (2013): 29–44.

Despret, Vinciane and Michel Meuret. "Cosmoecological Sheep and the Arts of Living on a Damaged Planet." *Environmental Humanities* 1, no. 8 (2016): 25–36. https://dx.doi.org/10.1215/22011919-3527704

Edge, Katelyn, Emma Johnston, Anthony Roach, and Amy Ringwood. "Indicators of Environmental Stress: Cellular Biomarkers and Reproductive Responses in the Sydney Rock Oyster (Saccostrea Glomerata)." *Ecotoxicology* 21, no. 5 (Jul 2012): 1415–25. https://dx.doi.org/10.1007/s10646-012-0895-2

Grosz, Elizabeth. *Chaos, Territory, Art: Deleuze and the Framing of the Earth*. New York: Columbia University Press, 2008, Kindle.

———. *Time Travels: Feminism, Nature, Power*. Crows Nest: Allen & Unwin, 2005.

Haraway, Donna. *Modest_Witness@Second_Millennium. Femaleman_Meets_Oncomouse: Feminism and Technoscience*. New York: Routledge, 1997.

———. "The Promises of Monsters: A Regenerative Politics for Inappropriate/D Others." In *Cultural Studies*, edited by Lawrence Grossberg, Cary Nelson, and Paula A. Treichler, 295–337. New York: Routledge, 1992.

———. "Staying with the Trouble for Multispecies Environmental Justice." *Dialogues in Human Geography* 8, no. 1 (2018): 102–05. https://dx.doi.org/10.1177/20438 20617739208

———. *When Species Meet*. Minneapolis: University of Minnesota Press, 2008.

Hustak, Carla and Natasha Myers. "Involutionary Momentum: Affective Ecologies and the Sciences of Plant/Insect Encounters." *Differences: A Journal of Femininst Cultural Studies* 25, no. 5 (2012): 74–118. https://dx.doi.org/10.1215/10407391-1892907

Ingold, Tim. *The Life of Lines*. Oxon and New York: Routledge, 2015.

Judith, Kate. "How Mangroves Story: On Being a Filter Feeder." *Swamphen: A Journal of Cultural Ecology* 7 (2020). https://openjournals.library.sydney.edu.au/index.php/Swamphen/article/view/14366

Kingdom, Amorina. "Creatures with Circatidal Rhythms Move to a Different Beat." *Hakai Magazine*. 2020. www.hakaimagazine.com/news/creatures-with-circatidal-rhythms-move-to-a-different-beat/

Kirby, Vicki. "Originary Humanicity: Locating Anthropos." *philoSOPHIA* 8, no. 1 (2018): 43–60.

Marder, Michael. "Plant Intentionality and the Phenomenological Framework of Plant Intelligence." *Plant Signaling and Behavior* 7, no. 11 (Nov 2012): 1365–72. https://dx.doi.org/10.4161/psb.21954

———. *Plant-Thinking: A Philosophy of Vegetal Life*. New York: Columbia University Press, 2013.

Massumi, Brian. *Parables of the Virtual: Movement, Affect, Sensation*. Durham and London: Duke University Press, 2002, Kindle.

———. *Politics of Affect*. Cambridge and Maldon: Polity Press, 2015.

Mensch, James. "Arousal: The Intertwining of the Within and the Without." *World Picture* 4 (2010). www.worldpicturejournal.com/WP_4/Mensch.html

Merleau-Ponty, Maurice. *The Phenomenology of Perception*. Translated by Donald A. Landes. Oxon and New York: Routledge, 2012, Kindle. Originally published as *Phénoménologie de la Perception*, Editions Gallimard, Paris, 1945.

———. *The Visible and the Invisible* Evanston: Northwest University Press, 1968. Originally published as *Le Visible et L'Invisible*, Editions Gallimard, Paris, 1964.

Nassar, Delia and Margart Barbour. "Rooted." *Aeon Ideas*. 16 October 2019. https://aeon.co/essays/what-can-an-embodied-history-of-trees-teach-us-about-life

O'Malley, Maureen A. "The Ecological Virus." *Studies in History and Philosophy of Biological and Biomedical Sciences* 59 (2018): 71–79. https://dx.doi.org/10.1016/j.shpsc.2016.02.012

Trewavas, Tony. "Plant Intelligence: An Overview." *BioScience* 66, no. 7 (2016): 542–51. https://dx.doi.org/10.1093/biosci/biw048

van Dooren, Thom. *Flight Ways: Life and Loss at the Edge of Extinction*. New York: Columbia University Press, 2014, Kindle.

von Uexküll, Jacob. "A Stroll through the Worlds of Animals and Men: A Picture Book of Invisible Worlds." In *Instinctive Behaviour: The Development of a Modern Concept*, edited and translated by Claire H. Schiller, 5–80. New York: International Universities Press, Inc., 1934.

Wilcockson, David and Lin Zhang. "Circatidal Clocks." *Current Biology* 18, no. 17 (2008): R753–R55. https://dx.doi.org/10.1016/j.cub.2008.06.041

Wylie, John. "Depths and Folds: On Landscape and the Gazing Subject." *Environment and Planning: Society and Space* 24, no. 4 (2006): 519–35. https://dx.doi.org/10.1068/d380t

3 Ecological meaningfulness and the negotiation of criteria

All semiotic ontologies have an interest in meaning. They are explorations of how ontological standing happens through a coming-into-meaningfulness. As Susan Petrilli argues, where there is semiosis, there is meaning and value.[1] Semiotic reference occurs to the extent that it is relevant; that is of import or value within some referential context, which is, in turn, written from the continued intertanglings of semiotic references.[2] Value and meaning are enfoldings within semiotic processes.[3] While both biosemiotics and the semiotic materialisms of Karen Barad, Donna Haraway, Vicki Kirby, Gilles Deleuze, and Brian Massumi frame ontologies of material semiotic meaning, they do this slightly differently. In a perspective shift from levels to folds, from preferencing the immobile and regular to preferencing motility and disturbance, the focus used for seeking meaning changes. In the former, meaning is in some sense tethered to questions of consistency and narrative closure. In the latter, meaning is always in question, on the move within processes, transient and multi-perspectival. Approaches that move to folding release meaning from being tied to stasis and continuity or to a particular narrative perspective. In these ontologies, meaning as such is nowhere precisely identifiable, but everywhere there are processes of meaning-seeking between. There is a subtle but significant difference here between those material semiotic positions which require scalar differentiations as foundational, and those which mute these as effects within immanent processes of becoming. This chapter reviews these positions and strikes a path between, towards an account of meaning which accommodates the shimmer of indeterminacy.

Ecological meaningfulness

Oysters telling mangrove stories, mangroves telling oyster stories

Where should we look for the meaning of an oyster? Does the task of deciding what "to mean" means comprise asking questions about determining what the boundaries of a thing are? This might be answered in many ways, for example, determining the border of the oyster's bodily coherence,

DOI: 10.4324/9781003286493-5

or of its duration of being alive, or of the duration and coherence of the consistent genetic heritage that a family or species shares. Or does it mean finding a narrative into which the oyster's connections with other things contribute? Again, this would be answerable in numerous ways, from "mangrove ecosystem" to "the evolution of molluscs" to "dinner". Or does it mean looking for a narrative of becoming, the oyster as an event in time and space tellable as a complex story of material semiotic processes, locating meaning with the processes of oystering? This question also opens up to multitudes of choices such as where to begin, how granular to delve, how to decide upon exclusions and inclusions, and what to do with everything unable to be included that nonetheless impacts. Or is it about asking about the oyster's impacts? These might range from providing food for a whelk, to the composition of the estuarine mud, to the role its future descendants might play in mitigating sea level rise. Again, there are many choices about where to finish following the story and how widely to range in scope.

All of these possible ways to approach what meaning means involve carving a narrative line through an ecological semiosis and they all hang upon there being a transcendent level from which that narrative thread can be identified as coherent. These questions already imply that there is something that makes the choices the questions imply, and whatever is in the position to have those choices determines the meaningfulness of the oyster. In these kinds of questions, the oyster is meaningful for something that throws a web of semiosis over the world and carves narratives across it – to a subject – which, within anthropocentric paradigms, is understood as a human within a group of humans entangled in common epistemological and narrative structures and processes.

If meaning is understood beyond anthropocentrism as semiotic material, or semio-ecological, the character of meaning changes dramatically compared with tying meaning to a human orientation. Understood beyond a human-centred frame, many things enfold an oyster within material meaningfulness. How the mangrove responds to the oyster's mud-sorting and current-altering enacts what the oyster means for the mangrove. The oyster is involved in performing the composition of the mud and water through its filter-feeding, physically sorting the mud particles and changing them into either oyster body, energy, or discarded items, by either digesting them or coating them in mucus and bacteria so they flocculate from the water current. This activity makes a foundational difference to the biological and geological composition of the water and the mangrove mud. Wendy Wheeler and Louise Westing write:

> Comparing and interpreting are, of course, relational, semiotic activities. It is precisely these activities of information and energy transfer between organism and environment that, in the semiotic activity of comparing beneficial and non-beneficial aspects of the world, and by that very act of doing, constitute both semiotic object world and semiotic subject.[4]

The complex oyster physiology of eating and excreting are the semiotic material processes the oyster performs (its reading, interpreting and writing we could say) as it contributes to making the mangrove and the shoreline. In making that difference, the oyster matters for the mangrove.

Some semiotic materialists, notably Haraway, would describe the emergent processes of becoming of the mangrove as mangrove storying.[5] We can say that oysters are meaningful within the storying of the mangrove.[6] Furthermore, storying between the oyster and the mangrove is both-ways criss-crossing. As the oyster is meaningful within the story of the mangrove, the mangrove is also meaningful within the material semiotic story of the oyster. As discussed in Chapter 2, in relation to the concept of synecdochy, the oyster is as made of mangrove as mangrove is made of oyster. In thinking like this, it is clear that very many stories of the meaning of that oyster are being materially performed by all the varied living and non-living processes that include the oyster. No Umwelt[7] occupies a preferential position with regards to the meaning of the oyster. What an oyster means will be multiple and widely varied. Ecological meaningfulness is wildly meaningful – full of diverse readings, interpretations and co-becomings.

Ecological criteria and obligation

Ecological systematicity involves more than just many different material stories. There are shared meanings across the stories. They overlap, enabling understandings between organisms that are very different (these kinds of understandings will be explored in more detail in Part 2). One way to think about the development of the shared meanings that make an ecological system is through focussing upon the development and establishment of criteria of differentiation. In comparing and sorting beneficial and non-beneficial aspects of the world according to its own estimations, an oyster is actively creating the criteria differentiating mudbank from water. The semiotic criteria for sorting shore from water are continually being practiced and refined through and within the bodies of oysters. A fold in the world, where tidal rhythms meet weathering shores, eventually comes to divide oyster from tidal flow, and through the oyster, to determine nutritious mudbank from murky water. Once there is difference, it can become a question – difference asks: "of what does this difference comprise and how does it come about?" This was the intent of Deleuze's concept of the refrain, introduced in Chapter 2. An oyster body is both the question and its response. It is also the ongoing refinement of the criteria for the meaningfulness of the question. As other organisms identify the mudbank as meaningful – the mangrove propagule which settles upon it, the crab or egret who live upon it – the criteria differentiating mudbank from current are performed in increasingly diverse ways. The difference becomes material-semiotically firmer as a wider range of varied indicators develops for it. Indicators may be as varied as the flight paths of insects, the movements of crabs and fish, changes in temperature, or the capacities of mangrove trees. A wider range of indicators invites an ever-wider range of interpreters to be able to identify that particular difference and to share its meaning albeit through the application of widely varied criteria.

Ecological commitments to meanings become normative as emergent meanings become more widely shared. As humans, we are familiar with the way linguistic meanings become normatively conventionalised through usage. In material semiotics, oyster bodies, cities, practices, mangroves – patterns that come to have a hold on relations for some time – do so as they make a knot of shared understandings that establish something like material norms (or habits in Peircean terms, as discussed in Chapter 1). Temporal and spatial material semiotic habits can become a sorting mechanism. What is grasped as meaningful becomes an attractor. A few mangrove trees establish and provide shelter for fish, and more fish and cormorants come to join, bats visit the flowers, humans gather oysters. Material habits start to make material normative claims possible. Mangal relations, for example, can be claimed as such, as opposed to claims of saltmarsh or seagrass, taking on the particular relations and processes that normatively (that is, as a result of the increasing commitment towards that habit) signal an ecosystem for those within it and those who interact with it. Likewise, certain filter-feeding decisions trend towards the developing material habits of oysters (decision is discussed further in Chapter 8). Trajectories align with affordances, becoming meaningful within their parameters, as eater, as vector, as oyster, and in doing so become differently from how they might otherwise, forgoing alternatives.[8] Becoming meaningful suggests normative trends and obligations which constrain as well as enable.

Meanings can be obligated to greater or lesser extents, with more or less rigidity. Perhaps we can say meaning can be more or less narratively pliable or have varied degrees of narrative commitment. Ontologies that regard meaning as more rigid offer value parameters and ethical frameworks interpretable as givens, such as rights or principles. Ontologies that shift the focus to the thick in-betweens of indeterminacy loosen the threads of these obligations and put the emphasis upon the up-close interactions in-between, asking for attentive inter-relational care, for example,[9] or more broadly for attentiveness and vulnerability within the both-ways-between.[10] Both approaches involve meaning with obligation, but the focus is upon different kinds of obligations, and what is of concern changes.[11] The question of obligations within entangled semiotic material stories will be developed further in Chapters 4 and 6 within a discussion of narrative strategies to support pliable and attentive ways of relating. How meaning differs between a layered or folded ontology is discussed next.

Meaning and folds

Meaning across levels or within folds

In biosemiotics, biological meaningfulness is accommodated in a hierarchy of levels. Autopoietic or self-reflexive processes round off bits of biological processes into cells, organisms, species, and communities and these material semiotic levels of relative abstraction make meaning possible as something at one level becomes meaningful to something at a level from which it can be discerned. This occurs as the meaningful concept emerges as contextually situated within a particular semiotic web located at that more semiotically abstract level. Wheeler, for example, argues that levels with gaps between them are needed to preserve continuity and to hold onto a circularity of reflexivity.[12] Biological life comes to mean individual animals, species, and communities, and the values associated with meaning converge around maintaining the integrity of bounded biological entities. This is not at all to say that these biosemioticians downplay process, relationality, or semiotic contextuality. They certainly do not. This is about where semiotic meaningfulness comes into play within their models. While the material world is constantly interacting, for these thinkers, meaning happens with the reflexive turn that enacts a gap between levels across which semiosis occurs.

Building this structure rigidifies ontological insights, it arrests biology, to borrow Massumi's term, into individuals, species, communities, and ecosystems. What goes on in between, all the messy responding and experimenting, becomes marginalised. As Massumi puts it:

> to the extent that the in-between is conceived as a space of interaction of already-constituted individuals and societies, middle-feeders end up back on the positional map. The tendency is to describe the in-between as a blending or parody of the always-already positioned.[13]

Finding meaning in the settling of fluidity into levels privileges stable forms that emerge from relational interaction and associates meaning with what coheres, with moments of stasis within the broader context of relational becoming. In aligning meaningfulness with coherence, biosemiotics is in alignment with a model of ecosystems science that locates its functional goals in such concepts as sustainability, adaptive capacity, and resilience, concepts defining the continued coherence of an ecological system. Critiques of ontological models that locate the meaning of an organism as its boundaries, whether as its duration or in space as its extent, are widespread. Stephen Shaviro, for example, points out how many accounts of life fail to adequately account for change and variety:

> Even the alternatives to the neo-Darwinian synthesis that are sometimes proposed today […] share mainstream biology's overriding

> concern with the ways that organisms maintain homeostatic equilibrium in relation to their environment and strive to perpetuate themselves through reproduction. It would seem that organic beings only innovate when they are absolutely compelled to, and as it were in spite of themselves.[14]

With ontologies of movement the focus changes from moments of stasis that can be identified within processes, to the processes themselves. What was marginal becomes the centre. By material semiotic accounts such as those of Kirby or Barad, meaning is identical with the communicability of materiality, with how inter-responsiveness registers inter-affect across becoming difference. This is a world of multiple perspectives where resolution is a narrative effect of contextual attunement across the interfaces of particular affective processes, always playing the edges of indeterminacy. Meaning ceases to be the gateway to a reality out there, meaning itself gets "demoted and treated as the effects of translations", as John Law describes in reference to Actor Network Theory and Deleuzian logic.[15] Law summarises; "there is no overall social, natural or conceptual framework or scale within which events take place: as webs grow they tend to grow their own metrics".[16] For Massumi, writing against rigid hierarchies of abstraction as the organisation of meaning, this space between determinacy and indeterminacy is indeed the world, and meaning, to the extent that it can be thought of at all, remains unstable, flickering in and out of perceptual potential. This stance seeks to hold open the sense that meaning is forever in-process.

As meaning comes to be found in the work of material narrativising, its practice comes to the fore. With a layered ontology, meaning can be located within a potentially knowable landscape of scalar strata, enabling navigation and simplifying or downplaying the performative activities of affective materiality. But, as Anna Tsing points out: "scale is not just a neutral frame for viewing the world; scale must be brought into being: proposed, practiced, and evaded, as well as taken for granted. Scales are claimed and contested in cultural and political projects".[17] Tsing is writing about human contexts, but scales also emerge through frictions of inter-responsivity beyond the human context. Semiotic scales of abstraction are practiced, and that practice involves exclusions, choices, and experiments. Moving from nested layers to infinite folds, the details of these practices become the concern. It is not the oyster alone whose agency in creating the categories of filtration produces meaning upon its materiality. The oyster only does this within the relationships it is enfolded into and that are enfolded into it, the constraints and affordances these relationships offer it, and the performances of becoming meaningful to each other that they are participating in. By this account the oyster means all the in-betweens it is entangled within, and a politics of mutually responsive taking-into-consideration is folded within each ongoing relation.

Meaning and virtuality in an immanent world

The meaningfulness of emergent identities, relations, and boundaries poses a challenge for thinkers such as Deleuze, Massumi, and Barad who insist upon remaining with a folded immanence. They accommodate such emergence with difficulty, as a constant effort of writing into the question. Deleuze and Massumi pursue this challenge by asserting the primary importance of the processual between, insisting that processes of enfoldedness, interactivity, and recursivity are more fundamental and hence more theoretically important than the abstractions that emerge from them. Things and relations never quite emerge but are active, motile, and experimental. Deleuze and Guattari admit territorialities occur but note these "would be nothing without these movements that deposit them".[18] Massumi presents accounts that include the emergence of levels of abstraction (he uses a soccer game as an example),[19] but against this insists upon "the superiority of the analogue".[20] Asserting the analogue or the virtual as the superior partner never entirely removes the challenge, but repositions it as a constant ingredient of processes that are always working the between. Massumi argues further that positions that value arrestedness over motion have indeed made an error. He claims that those who take emergence to be an outcome are erroneously claiming there exists a still world of standing patterns.[21] Any apparent state of stasis, he argues, only occurs because it is held in place through ongoing work and activity, and (to borrow Isabelle Stenger's term), through the obligated orientations of those processes turned towards its performance.

Deleuze's concept of the virtual carries a significant explanatory load within his orientation towards immanence. As Massumi takes it into his work, the virtual is an opaque world of potential, the site of the "bubbling of structuration in a turbulent soup of regions of swirling potential"[22] from which emergences may arise or actualise, when there is the capacity to discern them. It is that potentiality that is available for the diverse mechanisms of perception, discernment and story/meaning-making to interact with. As "the pressing crowd of incipiencies and tendencies",[23] it is pre-loaded with the stuff of the world that will, later, be found to have been in process towards the emergent outcome. Virtuality acknowledges that any "before" is already loaded with the pattern that will later be known to have been needed to unfurl the event of emergence or actualisation. By this account, emergence occurs because there is some compelling force of virtuality within the immanent domain of jittery intra-affect. But without discernment, it remains virtual, only actualisable in the world if there is the means to make it so. Playing a similar role for Massumi as the void of quantum field theory for Barad,[24] virtuality offers a source of unresolved potential. Massumi's ontology locates the becoming of a meaningful semiotic materiality as an interpretive interface that is always at play both ways across the virtual-actual or immanent-emergent.

Virtuality is enfolded into oyster perception as it discriminates the tidal particle load in ways relevant to its Umwelt. It can only do this because its perception and capacity to discriminate are already attuned within a broader patterning that supports such estimation. It has a previous familiarity, both entangled within its phenotype and accumulated through experience, of current, tidal patterns, the qualities of particles, and what counts as edible and inedible. The tidal current it discerns is not, however, an invention of its perceptual cognitive apparatus. There is a virtual-material logic to the tidal landscape that enfolds with its discernment. Activity was occurring immanently, both within oyster and within the regular movements of water currents, such that, upon the folding of the two together with the process of filter-feeding, each particular interaction between flesh and current might emerge. Oyster perception and discrimination are therefore not apart from virtuality. They are folded in, jittering also in negotiation, intra-affected within. Perceptual assemblages have been on their own journeys towards becoming capable of encountering the patterns affecting them. The move from immanence to emergence is not, therefore, a sudden revelation out of nowhere. There is no stepping up and down across discrete levels because perception is already there within the virtual, in the process of becoming along with the things it will find meaningful, it is part of the same (none of this implies oyster perceptions are somehow objectively correct – that is not the concern here).

In the river estuaries, the rhythm of the tides, the pulsing offerings of nutrients, the shape of the coast, and a particular evolutionary history of genetic development combine to offer the potential for becoming-narrativised into many bodily and more-than-bodily materialities. A theory of virtuality suggests those materialities are virtually present before they make their actual appearance. Applying an oyster semiotics, virtuality becomes a relational concept, associated with a particular material narrative of becoming, in this case an oyster-relative narrative. The tidal flux carries innumerable particles of sand, silt, plastics, fragments of algae, protozoa, bone, shell, wood, and more, and each of these is densely populated with viral and bacterial life. The material storying of filter-feeding oysters renders these into oyster flesh, clear water, and mudbank. The unfiltered flow is virtual in relation to the filtering and sorting process. The unfiltered stream is not inaccessible or unformed, it is just not yet oyster or mudbank. In a future without filter-feeders, this virtuality would not be relevant. Tidal wash would be merely that.

Perspectives backfolding

This chapter is building a picture that finds meaning within processes of immanent becoming interfolded with interpretive processes, each continuing to inform the other. Massumi illustrates this in a discussion of the both-ways criss-crossing relationship between perception and emergence in the classic case study of the orchid and the bee. He explains that the flower itself, as it evolves in its species form and as it grows as an individual within its environment, is materially affected by every one of the ways it is interpreted (we could say "storied", to emphasise the semiotic entanglement of any interpretive act). "Every creature connecting with a flower will think-perceive it differently" depending upon that creature's needs, he writes.[25] It is performed through these thought-perceptions, as they all are folded back within inter-responsive material relations. The flower "is all of the thought-perceptions in which it is implicated".[26] The myriad interpretations of the flower fold back into the processes of its material becoming and also fold the flower out into the wider world. The flower exists within the interface between its immanent material becoming, which remains virtual until perceived, and all the many interpretations of it into actuality. In general: "Things, perception, and thought are in a reciprocal movement into and out of each other and themselves".[27]

Here becoming meaningful operates as a double-folding that resonates, for me, with the forwards-and-backwards washings of the tides. In the forwards sense, the band of mangroves is an emerging or assembling of the interactions of all the critters and non-living entities within it and all the impacts upon it. In the backwards sense, the mangrove band will be something different depending upon each experience of it. What it is for me as I walk along the riverbank will be different from what it is for the shark waiting at its edge for the tide to change and draw fish out from its protective cover, for the mangrove seedling settling to the mud as the current it was floating along on is quietened by the tree roots, or for the landowner dreaming of a sandy riverbank. As each of us changes our behaviour in response to them, all of these experiences fold back into affecting the mangrove. This backward sense of becoming is a coming after that which it generates. In this sense, it is only by having an impact, by being prehended or experienced that the affecting thing becomes. Forwards and backwards interplay, and mangrove becomes as complex negotiations between.

In a folded semiotic material ontology, the emergence of a concept into meaningfulness, to the extent that it occurs, is not a singular edge demarcating a change of state or a jump to a higher level.[28] It is quivering process, unfolding out and refolding back in, unstable and inter-folded upon itself, better described by verbs like emerging or becoming, and remaining in its verbal form. Likewise, abstraction does not occur as a singular act of destructive imposition upon a cooperative micro-relationship-based world. It ceases to be meaningful as a singular cut or leap and needs to be reconceived as a semiotic event occurring over time. At no single moment is there a sudden shift to a different logical order from the moment before. Relationships of relative abstraction across scales shimmer into and out of virtuality/actuality, in and out of being discerned, with no hierarchical qualification. As Massumi seeks to remain in the back-and-forth criss-crossing of the both-ways-interface, things and relations come into their relative and responsively changing forms within fields that flicker on the edge of virtuality. Perceiver and perceived are both unstable co-emergents of the same flickering. "Indeterminacy and determination, change and freeze-framing, go together. They are inseparable and always actually coincide while remaining disjunctive in their modes of reality".[29] Ontological levels of abstraction play out in entangled stories that continue to tell themselves, with the teller and hearer enfolded within the virtual meaningfulness. The teller and hearer of the story are inside-and-outside the story. We can see this easily in the way that language only exists in being performed and iterated, as the mangrove only exists in its trees, animals, and muddy processes, and as the oyster only exists in its communications with the tide and the shore – reflecting the synecdoche referred to in Chapter 2.

Shimmer, indeterminacy and the retreat from completion

Shimmer

Shimmer, iridescence, indeterminacy, jitter, and liveliness are among the various terms employed to describe the sense of creative instability found at the interface within what I have been referring to as immanence/emergence or forwards and backwards. These are concepts of a discourse of the between, that tries to figure out how to stay with the instability and proximity of the negative, with indeterminacy, negotiation, and experimentation there.

As Maurice Merleau-Ponty wrangles with the interface between being (as that which is perceived to positively exist) and negation (as the negation of the perceiver's perspective that is removed from consideration as the world is taken as positively existing), he describes the world as like a shimmering "band of foam" cast upon an ocean wave along this interface.[30] The "shimmering and living" effect is only visible when the wave is viewed from close up, where the dark watery negative and the foamy positive cannot be distinguished except as an ever-changing shimmer of between. From afar the band appears as a single immobile line. Merleau-Ponty's point is that only at a scale of relevant abstraction can anything in the world be discerned as positively existent, at a scale where the active enfoldedness of perception and perceptual apparatus is negated in an effect of distance. Those who make this move are considering from afar, a perspective that enables them to eliminate the evidence of the lively shimmer of interrelationality.

Deborah Bird Rose writes of shimmer within the context of Indigenous Australian cultural practice and more-than-human material co-becomings. Her usage initially might appear to have little in common with that of Merleau-Ponty; however, on closer consideration the term connotes something very similar for both. For Rose, shimmer (or iridescence) describes that which enlivens the world, within human, more-than-human and beyond-human contexts. Rose grounds her discussion upon Howard Morphy's anthropological study of the Yolgnu term bir'yun, roughly translated as "brilliant or shimmering".[31] Morphy's study explores bir'yun primarily as Yolgnu use it to describe the ways their paintings come to shimmer with the deep liveliness of the country as the artists incorporate their cultural and embodied knowledge of country into the work. Rose focuses in on the up-closeness of this shimmer, writing its unfolding process as it is danced up within Indigenous dance practice and as it is called up in the meeting of tongue and flower as flying-fox and eucalyptus blossom respond to each other. "It is the great, expressive, demonstrative 'yes' ", she exclaims, as the "lush, extravagant beauty, flamboyance, and dazzling seductiveness" of the blossom "reaches out into the world [...] and for their part, the flying-foxes come racing to respond".[32] Grey headed flying-foxes (*Pteropus poliocephalus*) have established colonies in small, dense patches of Eucalyptus trees located in moist gullies abutting mangroves on the Georges River and on Wolli Creek, a tributary of the Cooks River. The Wolli Creek colony sometimes expands to over 20,000 individuals and is recognised as a camp of National importance.[33] When the colony is large, the stream of foxes leaving their camp to forage on blossoms and fruit across the city during the evening fly-out is a remarkable experience for many local human inhabitants. We stand mesmerised and grateful for that great "yes" of shimmer.[34]

As in Merleau-Ponty's exploration of shimmer, Rose's discussion of bir'yun also entails an acknowledgement of the importance of an incorporation of the negative. Rose explains how shimmer is unstable. It pulses with an unsteady power derived from the rhythms of diverse ecological and cultural pulses, one important pulse being that of the tropical wet and dry seasons. The dry season mutes the shimmer, negating its becoming a source of reliable mastery. Shimmer is its lightness and also its darkness: "for power to come forth, it must recede. For shimmer to capture the eye, there must be absence of shimmer. To understand how absence brings forth, it must be understood not as lack but as potential".[35] Other rhythms diffract across the wet/dry pulse, so shimmer is tantalising but mysterious in its suggestions of predictability:

> the work of the sun ensures that rain does not take over the earth, and the work of rain ensures that the sun does not take over; their pulse is complementary, and each part arrives as relief from an excess of the other. There are winds, of course, and the play of winds on grasses and on clouds. The effect, if one is attentive, is an iridescence or shimmer sparking up at these larger scales.[36]

With shimmer the negative is the source of instability that provides potential for the positive. Shimmer honours the indeterminacy, in-betweenness, and interactivity of ecological liveliness.

Playing across the fold

In the worlds of becoming of Deleuze, Massumi, Haraway, Kirby, and Barad, discernment processes pull up short of a totalising meaning or emergence. Meanings are not completed, narratives do not have neat tied-up conclusions. Rather, meaning plays across that fold where perception steps outside and turns around and discerns the emergent, staying with the process such that as perceptions begin to pull coherencies out, as meaning approaches, senses reawaken and notice the new changes, further affects, new directionalities, other interpretations. Insights fold back, while yet unresolved, to filter through another time, and another. Meaningfulness jitters across that edge between immanent and emergent discernment. Layers, scales, and meaningfulness are thinkable, but barely and only as variable, complex, situated, and relational processes.[37]

In Massumi's discussions, instability and intensity are associated with a system's self-referentiality and lead to "hypermutable" states of criticality, from which transformation occurs in a self-reflexive folding move.[38] Barad locates this kind of troubled indeterminacy and doubled being/non-being of the positive and negative at the scale of the "quantised indeterminacies-in-action" of virtual particles.[39] These "ghostly non/existences that teeter on the edge of the infinitely thin blade between being and nonbeing"[40] are kin to the quantum vacuum, which fluctuates with "indeterminate vibrations".[41] Vibrations within quantum phenomena *are* the virtual particles/waves – what we call them depends upon our language paradigm and measuring apparatus. An electron is a particular measurable kind of instability and indeterminacy across the non-edge between the void and measurability. If our apparatus registers affect, we call the electron existent; how it is measured provides the parameters for what the electron is understood to be. Quantum field theory knows these measurements are picking up flickering unstable interactions.[42]

Oysters exist as a delicate in-between, becoming flesh in response to the regularities, risks, and materialities of the turbulent tidal zone. To my sensing apparatus an oyster is more certain and fleshier than a quantum field, but the actual oyster is particular and uncertain – a matter of chance and negotiation between what the tide provides and what a body can do.

This chapter has focused on what meaning entails within a semiotic materiality. It explored in particular the difference it makes to meaning whether the ontology, like biosemiotics, embraces categorical criteria that perform semiotic material layers, or whether these are rejected in favour of an infinitely interfolding immanence. Within a layered ontology, questions of meaning tend to be concerned with boundary coherence and relations construed as between agential actors or subjects. In moving to ontologies that recognise a folded immanence, questions of meaning become concerned with the entangled obligations and values that emerge through, and facilitate or detract from, attunements, irritations, challenges, and accommodations negotiated within the thick betweens. Similarly, while meaning and obligation are entangled within both ontological approaches, within layered ontologies obligation becomes concerned with boundary discourses of inclusions and exclusions, whereas in folded approaches, obligation discourses concern attentiveness, vulnerabilities, and sensitivities across interfaces.

Folding regards the interfaces across which semiotic interpretation occurs as lively sites of co-becoming, sites where same meets same across an intra-affective in-between where meaning becomes along with difference. With a folded semiotic materiality, interpretive processes bring difference, but not scalar separation. Folds are pliable, supporting the sense that emergence into discernment is a thick process with ambiguous indeterminate edges, rather than a divisive cut or jump. Rather than layers, folds form fractal multiplicities. Interpretive relations of co-becoming are everywhere, but situated and varied, rather than forming hierarchies. Meaning within this approach becomes concerned with issues of becoming together through interaction. It is forged through processes of attention, experimentation, and speculation. Obligation becomes concerned with the doing of material semiosis, with the processes of responsive folding, with attention towards how material meaning is made with others through co-affect. The three chapters of "Strange Reflexivities" have worked questions of being in-between through the rhythms of the tides as they temporalise and mark changes across the estuarine mangroves, and through the bodies of oysters as they become in response to tidal rhythms and in turn contribute to the muddy foundations of the mangrove.

Notes

1 Susan Petrilli writes, for example, that semiotic research: "inevitably relates signs and values, semiotics and axiology, signification and significance, meaning and sense, semantics and pragmatics", in *Sign Studies and Semioethics: Communication, Translation and Values* (Boston: De Gruyter, 2014), xix.
2 Petrilli, "Sign Studies", 7.
3 Ibid., 7–10.
4 Wendy Wheeler and Louise Westling, "Biosemiotics and Culture: Introduction," *Green Letters* 19, no. 3 (2015): 223, https://dx.doi.org/10.1080/14688417.2015.1078973.223
5 Donna Haraway, for example, writes: "Flesh and signifier, bodies and words, stories and worlds, these are joined in naturecultures", in *The Companion Species Manifesto* (Chicago: Prickly Paradigm Press, 2003), 20.
6 See also Kate Judith, "How Mangroves Story: On Being a Filter Feeder," *Swamphen: A Journal of Cultural Ecology* 7 (2020).
7 "Umwelt", is the term chosen and popularised by Jakob von Uexküll to describe the phenomenal world, the world as it is available to a particular experiencing frame of reference, that is, as it is sensible (in the broad sense referred to in the previous chapter of this book that would include plant senses, fungal senses, dispersed senses and so on) in "A Stroll through the Worlds of Animals

and Men: A Picture Book of Invisible Worlds", in *Instinctive Behaviour: The Development of a Modern Concept*, edited and translated by Claire H Schiller (International Universities Press, 1934).

8 Deacon's explorations of the constitutive significance of constraint consider this from a biosemiotic perspective. See Terrence W. Deacon and Tyrone Cashman, "Eliminativism, Complexity and Emergence", in *The Routledge Companion to Religion and Science*, ed. James W. Haag, Gregory Peterson, and Michael Spezio (Routledge, 2011), 193–205.

9 See, for example, María Puig de la Bellacasa, *Matters of Care: Speculative Ethics in More Than Human Worlds* (Minneapolis and London: University of Minnesota Press, 2017), Kindle.

10 This point will be discussed further in Chapters 4 and 6, particularly in relation to work by Rose, Stengers, and Haraway.

11 In reflecting upon the practices of science, Latour has argued that "matters of fact" are better described as "matters of concern", pointing out that what matters concerns semiotic relevance and suggesting that this shift might prompt more caring engagements, in "Why Has Critique Run out of Steam? From Matters of Fact to Matters of Concern," *Critical Inquiry* 30 (2004): 225–48.

12 Wendy Wheeler, "Gregory Bateson and Biosemiotics: Transcendence and Animism in the 21st Century." *Green Letters* 13, no. 1 (2010): 49. These associations of meaning with levels of abstraction within biological theory go back at least to Jacques Monod who asserts that the purpose, meaning, and value of an entity are relative to the "more comprehensive project" at a level of greater abstraction, within which the particular lesser-scaled project has meaning. "[I]t is only as a part of a more comprehensive project that each individual project, whatever it may be, has any meaning", he writes, in *Chance and Necessity*, trans. Austryn Wainhouse (New York: Random House, 1972), 14.

13 Brian Massumi, *Parables of the Virtual: Movement, Affect, Sensation* (Durham and London: Duke University Press, 2002), 69.

14 Steven Shaviro, *Without Criteria: Kant, Whitehead. Deleuze, and Aesthetics* (Cambridge. Massachusetts Institute of Technology Press, 2009), chapter 4, paragraph 35, Kindle.

15 John Law. "Actor Network Theory and Material Semiotics," in *The New Blackwell Companion to Social Theory*, ed. B. S. Turner (Oxford: Blackwell, 2009), 8.

16 Law, "Actor".

17 Anna Lowenhaupt Tsing, *Friction: An Ethnography of Global Connection* (Princeton and Oxford: Princeton University Press, 2005), 58.

18 Gilles Deleuze and Felix Guattari. *A Thousand Plateaus: Capitalism and Schizophrenia*, trans. Brian Massumi (Minneapolis and London: University of Minnesota Press, 1987), 55.

19 Massumi, *Parables*, 72–80.

20 Ibid., chapter 5.

21 Ibid., 7, 59.

22 Ibid., 34. Massumi uses this phrase in reference to Simondon's pre-individual, but this is in the context of Deleuzian ontology.

23 Ibid., 30.

24 Karen Barad, "Troubling Time/S and Ecologies of Nothingness: Re-Turning, Re-Membering, and Facing the Incalculable," *New Formations* 92 (2017), https://dx.doi.org/10.3898/newf:92.05.2017

25 Massumi, *Parables*, 92.
26 Ibid.
27 Ibid., 94.
28 Ibid., 8.
29 Ibid.
30 Maurice Merleau-Ponty, *The Visible and the Invisible*, ed. Claude Lefort, trans. Alphonso Lingis, (Evanston: Northwest University Press, 1968), 64.
31 Deborah Bird Rose, "Shimmer: When All You Love Is Being Trashed," in *Arts of Living on a Damaged Planet*, ed. Anna Tsing et al. (Minneapolis and London: University of Minnesota Press, 2017), paragraph 8, Kindle.
32 Rose, "Shimmer", paragraph 23.
33 Wolli Creek Preservation Society. "Grey-Headed Flying-Foxes". Last modified 9 May 2020, accessed 26 November 2020. www.wollicreek.org.au/plants_animals/flying_foxes
34 I have written about the value of shimmer as a heuristic for inspiring honouring, particularly in relation to the 2019–20 Australian forest fires in Kate Judith, "Errinundra Shimmer," in *Continent Aflame: Responses to an Australian Catastrophe*, ed. Pat Anderson et al. (Armadale: Palaver, 2020), 51–63.
35 Rose, "Shimmer", paragraph 12.
36 Ibid., note 8.
37 William Connolly, for example, explains:

> You can assert that one entity is more "complex" than another if the second required the first as a platform from which it could emerge. But to say that does not necessarily commit you either to a single line of complexity or to project a cosmic trend in which things express an inexorable tendency to move from lower to higher complexity,

in *Facing the Planetary: Entangled Humanism and the Politics of Swarming* (Durham and London: Duke University Press, 2017), chapter 4, Creativity, Extinction Events and Attachment to the World, paragraph 10, Kindle.
38 Massumi, *Parables*, 109–10.
39 Barad, "Troubling Time/S", 78.
40 Ibid.
41 Ibid., 77.
42 Ibid.; Vicki Kirby, *Quantum Anthropologies: Life at Large* (Durham and London: Duke University Press, 2011), Kindle.

References

Barad, Karen. "Troubling Time/S and Ecologies of Nothingness: Re-Turning, Re-Membering, and Facing the Incalculable." *New Formations* 92 (2017): 56–86. https://dx.doi.org/10.3898/newf:92.05.2017

Connolly, William. *Facing the Planetary: Entangled Humanism and the Politics of Swarming*. Durham and London: Duke University Press, 2017, Kindle.

Deacon, Terrence W. and Tyrone Cashman. "Eliminativism, Complexity and Emergence." In *The Routledge Companion to Religion and Science*, edited by

James W. Haag, Gregory Peterson, and Michael Spezio, 193–205. Routledge, 2011.

Deleuze, Gilles and Felix Guattari. *A Thousand Plateaus: Capitalism and Schizophrenia.* Translated by Brian Massumi. Minneapolis and London: University of Minnesota Press, 1987. Originally published as *Mille Plateaux: Capitalism et Schizophréne 2*, Les Éditions de Minuit, 1980.

Haraway, Donna. *The Companion Species Manifesto*. Chicago: Prickly Paradigm Press, 2003.

Judith, Kate. "Errinundra Shimmer." In *Continent Aflame: Responses to an Australian Catastrophe*, edited by Pat Anderson, Sally Gardiner, Paul James, et al, 95–97. Armadale: Palaver, 2020.

———. "How Mangroves Story: On Being a Filter Feeder." *Swamphen: A Journal of Cultural Ecology* 7 (2020). https://openjournals.library.sydney.edu.au/index.php/Swamphen/article/view/14366

Latour, Bruno. "Why Has Critique Run out of Steam? From Matters of Fact to Matters of Concern." *Critical Inquiry* 30 (2004): 225–48.

Law, John. "Actor Network Theory and Material Semiotics." In *The New Blackwell Companion to Social Theory*, edited by B. S. Turner, 141–158. Oxford: Blackwell, 2009: 141–158.

Massumi, Brian. *Parables of the Virtual: Movement, Affect, Sensation*. Durham and London: Duke University Press, 2002, Kindle.

Merleau-Ponty, Maurice. *The Visible and the Invisible*, edited by Claude Lefort. Translated by Alphonso Lingis. Evanston: Northwest University Press, 1968. Originally published as *Le Visible et L'Invisible*, Editions Gallimard, Paris, 1964.

Monod, Jacques. *Chance and Necessity*. Translated by Austryn Wainhouse. New York: Random House, 1972. Originally published as *Le Hasard et la Nécessité: Essai sur la Philosophie Naturelle de la Biologie Moderne*, Éditions du Seuil, 1970.

Petrilli, Susan. *Sign Studies and Semioethics: Communication, Translation and Values*. Boston: De Gruyter, 2014.

Puig de la Bellacasa, María. *Matters of Care: Speculative Ethics in More Than Human Worlds*. Minneapolis and London: University of Minnesota Press, 2017, Kindle.

Rose, Deborah Bird. "Shimmer: When All You Love Is Being Trashed." In *Arts of Living on a Damaged Planet*, edited by Anna Tsing, Heather Swanson, Elaine Gan, et al, 51–63. Minneapolis and London: University of Minnesota Press, 2017, Kindle.

Shaviro, Steven. *Without Criteria: Kant, Whitehead. Deleuze, and Aesthetics. Technologies of Lived Abstraction*. Cambridge: Massachusetts Institute of Technology Press, 2009, Kindle.

Stengers, Isabelle. *Cosmopolitics*. Translated by Robert Bononno, vol. 1. 2 vols. Minneapolis: University of Minnesota Press, 2010.

Tsing, Anna Lowenhaupt. *Friction: An Ethnography of Global Connection*. Princeton and Oxford: Princeton University Press, 2005, Kindle.

Wheeler, Wendy. "Gregory Bateson and Biosemiotics: Transcendence and Animism in the 21st Century." *Green Letters* 13, no. 1 (2010): 35–54. https://dx.doi.org/10.1080/14688417.2010.10589069

Wheeler, Wendy and Louise Westling. "Biosemiotics and Culture: Introduction." *Green Letters* 19, no. 3 (2015): 215–26. https://dx.doi.org/10.1080/14688417.2015.1078973

Wolli Creek Preservation Society. "Grey-Headed Flying-Foxes." Last modified 9 May 2020. Accessed 26 November 2020. www.wollicreek.org.au/plants_animals/flying_foxes

Part 2

Monstrous relations

Exploring a hermeneutic account of relationality[1]

Thinking interstices within a material semiotic context calls for dwelling with the uncertainties and tensions of relations. Part 2: "Monstrous relations" follows the events occurring within the Georges River valley between a wallaby, a female saltmarsh mosquito (*Aedes vigilax*), the Ross River Virus (RRV) she carries, and myself, exploring concepts of relationality as they apply within our multispecies interactions. The three chapters of this section consider the following three issues: what relationality is as it is conceptualised within a semiotic materialism; questions relating to the ethical concerns raised by relations of ecological obligation; and what might be needed if the hope of finding an ethicality from a relational ontology is to have its potential redeemed following criticisms of an inattention to processes of privilege and exclusion.

Note

1 Occasional passages scattered through Part 2 have been published in Natalie Marr, Mirjami Lantto, Maia Larsen, Kate Judith, Sage Brice, Jessica Phoenix, Catherine Oliver, Olivia Mason, and Sarah Thomas, "Sharing the Field: Reflections of More-Than-Human Field/Work Encounters," *GeoHumanities* (2022), https//doi.org/10:1080/2373566X.2021.2016467

DOI: 10.4324/9781003286493-6

4 Lines of desire and knots of obligation

Animals with considerable psychological, social, and biological affect for humans emerge from the Georges River mangroves. Mosquitoes breed in the puddles of brackish water left in the higher reaches of the mudbanks after king tides. Some female mosquitoes pay particular attention to mammalian skin and blood, and humans have become especially attentive to and knowledgeable about these. Our bodies pay close attention to each other such that, not only our physiologies, but also our daily practices have become entangled. This chapter is grounded in the materiality of mosquito senses and behaviours. It discusses a female *Aedes vigilax* mosquito's search of a blood meal to explore a semiotic material account of relationality.

Relational ontologies

Trends towards relational ontologies occurred within many discipline areas through the twentieth century, often with the intention of correcting perceived injustices associated with categorical approaches. Val Plumwood and Donna Haraway[1] are among those who have criticised as hegemonic, ontologies that identify entities by their boundaries and assume difference as structural. Overly differentiated categories, they argue, become established within political, cultural, economic, and semiotic institutions and practices and take on differential values through processes of normalisation. Because of these processes, supporters of relational approaches have claimed that non-relational ontologies lead to unjust discrimination when they support the categories and values of privileged perspectives. In contrast, relational ontologies locate differentiation within processes. I draw upon three of the disciplinary heritages of relational ontology. First, anthropological studies increasingly recognised the relational basis of the ontologies of many societies and had been calling for epistemological methodologies that could value these on their own terms. I will be referring to the anthropological work of Elizabeth Povinelli, Deborah Bird Rose, Alberto Corsìn Jiménez, and Eduardo Viveiros de Castro. Second, within feminist theory, work on a care-based ethics had been growing since Carol Gilligan's research on gendered ethical identity.[2] This work, that on the interconnectedness and

DOI: 10.4324/9781003286493-7

gestationality of bodies,[3] and ecofeminist discussions of the connectedness of women and nature,[4] all contributed towards a strong focus from feminist theory into relational ontologies. Haraway's contributions within this tradition are notable and will be considered here. Third, Bruno Latour, John Law, Vinciane Despret, Haraway, and others with an interest in the culture of science and technology, organisational studies, ecology, and sociology have worked their way to relational ontology from analysing how outcomes, categories, identities, and worldviews emerge from the relations of humans in particular groups and the relations among more-than-humans, and I also draw upon this work. I focus particularly on two of these writers, Donna Haraway and Deborah Bird Rose. These authors both locate and maintain relationality at the core of their positions. Although each of them works within different contexts and concepts, I suggest they both describe semiotic worlds which are written and read within situations, from points of view, along trails of unfolding narrativity. My understanding of relationality takes shape within such a semiotic materiality, and this is why I have chosen to focus on their work. I note also that Rose wrote some of her later works from her home near the Georges River, entangling further the stories of site and theory.

Within this literature, relationality is often taken as something akin to a methodological approach which can enable work in a wide range of academic contexts – including pedagogical, psychological, political, and scientific – to better engage with the world compared with a more object-based boundary-focussed approach.[5] Relational methodologies privilege process over reification, holding things and events to be emergent from processes.

> By tracing the contextual and contingent paths along which such forms come into being, as opposed to populating the categorical spaces of assorted dualist narratives, relational thinking shifts our analytical focus to the ways in which entities, thought as processes rather than existents, become entwined.

is a fairly standard description offered by François Dépelteau.[6] I consider relational ontology and semiotic materiality to be interconnected.[7] Bryan Bannon, taking inspiration from Merleau-Ponty's phenomenology, helps explain the connection. He writes:

> From the perspective of a relational ontology in which properties are the result of a relationship (e.g. color is a relation between a body, light, and a visual organ), one must view continuity as deriving from the relations between beings rather than in some inherent property that each possesses.[8]

Since continuity is the result of relations, time and space occur only in the unfurling of sequences of affect and response, which are the building blocks of a material semiotics. For Bannon, "nature is a process of temporalization and spatialization that occurs between beings, which are themselves events, by virtue of their participation in an enveloping event".[9] "Nature" here stands for that which exists as meaningful. In other words, a meaningful relational material ontology is a semiotic one.

Lines of desire

Mosquito desire

To the landward side of the mangroves along the George's estuaries, protected from most of the tidal disturbances, brackish water forms stagnant pools. In the summer in Australia, mosquitoes are seeking pools of still water to lay their eggs. *Aedes vigilax*, the saltmarsh mosquito, breeds specifically along the edges of mangrove zones, and is abundant along the Georges River during the summer months. This species is known as a particularly savage biter[10] and is also the main vector for RRV throughout Australia. Indeed, the Georges River catchment is known as a hot spot for this virus.[11] RRV is the most prevalent mosquito-borne disease affecting humans in Australia, with around 5000 cases recorded annually across the country, although researchers believe the incidence of infection is probably greatly underreported.[12] The symptoms, flue-like arthritic pain, fatigue, and rash, can be severe or mild and also can last from a few weeks to several years.[13] Chronic joint pain lasting for years occurs for approximately 50% of sufferers and the symptoms can be extremely painful and disabling, resulting in "severe lifestyle constraints and substantial economic losses".[14] In the Georges River catchment, Australian native macropod species, particularly wallabies, are likely to be the main reservoir hosts.[15]

Studies on horses[16] and lab mice[17] find they suffer a range of RRV symptoms, including swollen joints and lethargy, so perhaps it is a stiff and weary wallaby who ventures out at dusk from the scrub into a grassy clearing near the river, for RRV has been multiplying rapidly within her blood for the past few days. Like other viruses, RRV consists of some replicating material in the form of strands of nucleic acids usually as RNA or DNA, and a coating of proteins. Because of this pared back simplicity, the virus requires the enzymes, ribosomes, and other cellular contents of an animal host cell to synthesise proteins, reproduce, and spread.[18] During the late stages of cellular infection, the virus reconfigures the mechanisms of its host cell so the cell produces large amounts of viral RNA and proteins instead of performing its other functions. Cells thus transformed rupture or disintegrate once the virus replication is completed, stressing the health of the host and releasing large amounts of virus into the animal's blood. To continue surviving and replicating, a virus needs a means for spreading between hosts. RRV, like dengue, chikungunya, and zika, is usually spread to humans via mosquitoes.

This same evening, a female *Aedes vigilax* mosquito crawling only days ago from a green-slimed drainage ditch alongside the mangroves of Lime Kiln Bay, launches out from the underside of the leaf that had protected her from daytime predators and the hot summer sun. She is hungry for some animal blood because she is pregnant, and her eggs require that blood for their development. The blood does not nourish the mosquito. Her nourishment comes only from plant nectar, as it did before she was pregnant. It is only the developing eggs that will use that blood. Already we cannot tell whose hunger this is.[19] Is her hunger her own, or does it call down through her cells from others? Is it a call from a future to live into its already diverging virtuality? Perhaps her hunger belongs also to her microbiotic guests, demanding her hospitality.[20] These guests include long-term bacterial and protozoan residents who contribute digestive products throughout her body and interact with viruses there. Our relationships are beginning, and already it is uncertain just who is becoming entangled and what their motivations may be. The mosquito body is interested in solutions rather than causes, she responds to all these hungers through the one corporeal pathway. She has one way to sate these multiple desires.

Following Gilles Deleuze and Felix Guattari, desire implies neither intentionality nor lack – it is rather the emergent being itself, produced through the entangled efforts and collective orientations of affect,[21] response, and negotiation among those who comprise a life form. Deleuze and Guattari propose that desire arises within a collective, as that version of a future around which the collective curls itself. They rework Antonin Artaud's concept, the Body Without Organs, as "that which one desires and by which one desires",[22] a concept that seeks to hold together as one the emergent being of a desiring body, and the disparate yearnings towards it. That desire towards and of the Body Without Organs is what draws her/them together. Ontologically, it replaces being with desire; mosquito being replaced with mosquito desire. This gives desire "a fascist determination".[23] It makes a tendency towards, a line of flight, a becoming circled around. It brings (or is) some degree – sometimes a profoundly directive one, other times a loose suggestion – of obligation, a directive that begs following. But "[d]esire is always assembled; it is what the assemblage determines it to be".[24] That obligation is maintained only by the determination towards it of those harnessed to its pleasures.

Deleuze and Guattari's approach to desire plunges us into the interfolded mutualities of emergent co-becomings. Any sense of continuity is the result of numerous relations, taking their particular times and courses, aligning with particular emergent desires.[25] A becoming being, such as this mosquito, will be both particular to and entirely of its circumstances, and yet always an approximation, a "having a go at" a rounding off or autopoiesis, in a form that can be described, this evening, as desire and the search for its satisfaction. This hunger for blood, and the physiologies, chemistries and behaviours called up in responding to its demands, asserts itself as a mosquito. This particular little body commandeers a hunger as being of it, not experiencing this as partial, multiple, or dispersed. Deleuze and Guattari will say, she is a little machine,[26] a desire-producing machine tuned forward towards the scent of warm mammals.

This mosquito hunger piques the sensitivity of many scent receptors, which can discriminate favoured warm animals by the particular mix of their many different odours, including their exhaled carbon dioxide.[27] She moves into the breeze, towards these odours. She hones in further when she sees the shape and size of a mammal, and still further as she detects the heat from skin.[28] Although RRV may have been present within the wallaby for some time, the period within which it is abundant enough within her blood that it may be picked up by a mosquito, the period of viremia, is only very short, from one to six days.[29] Today, huge numbers of RRV viral envelopes are circulating within this wallaby's blood. When the mosquito bites the wallaby, she drinks up a good dose of RRV.

Figured as a line of desire, a mosquito ceases to be a collection of organs inside an exoskeleton, a blob enclosed in a pre-given space and time. Instead, by this way of thinking, the mosquito is a particular unfurling of a pattern. She is a rolling forward of entangled trajectories into a story or perhaps a melody; continuities of material semiotic affect-and-prehension, question, and response, that remember their pasts into the spaces and times of their unfurling.

Entangling mangrove trajectories

If an *Avicenna marina* propagule can root itself and establish, if there is sufficient warmth, a source of mud, and enough fresh water among the salty, from rain or river, it will grow, and begin gathering about itself a complex mangrove world: "once mangroves are established they tend to further accumulate sediment, reducing depths and increasing the area suitable for the establishment of mangroves".[30] Mud dwellers such as crabs, worms, lobsters, and snails move in and wading birds arrive to feed from these. Fish find refuge among the establishing mangrove roots, and bigger fish and fishing birds such as cormorants lurk along the edges. In addition, tree canopies with nectar rich flowers provide food and shelter for still more inhabitants including flying-foxes, and pools of brackish water in mud puddles among the mangrove roots in the higher tidal areas offer opportunities for mosquito and other larvae to grow.[31] These then provide food for insectivorous birds, bats, and reptiles. My blood is carried within the bodies of mosquitoes into the mangrove, where it nourishes the next generation of mosquito larvae.

All these beings and things orient their lives and trajectories together to become the mangrove. But becoming mangrove does not mean these things cease to be also other than mangrove. While tangling or knotting together into becoming mangrove, each plant, algae, animal, mineral particle, and so on is also traceable along its own particular trajectory. The shark swims in from its journey down the coast, the mangrove seedling washes in on the incoming tide, I walk in and out along the path. We each are a material story writable along our own line – each of which is also comprised of tangled multiples of other lines. In thinking of co-becoming as entanglements of storied lines interweaving within a semiotic materiality, I am thinking of lines as a particular kind of discernible continuity of pattern of matter, an ongoingness along a directionality, and also a particular reading of that pattern. A line is a patterning that both can be, and is, read as continuing along in a directionality, as unfurling along.

Living along semiotic lines

Tim Ingold describes organisms as "bundles of interwoven lines of growth and movement, together constituting a meshwork in fluid space. The environment [...] a zone of entanglement".[32] Things in general, not only organisms, he writes "carry on along their lines". They are "goings-on", "entanglements".[33] Not only things, but also relations, are lines for Ingold. The relation is lived along – a being is active in its relating, relating is made by doing things and making sense of what is being done; not a between, but an along. Ingold is careful to point out that not only are things, organisms and relations themselves lines of unfurling becoming, so also is the activity of reading or interpreting or discerning them. He introduces this idea using the physical act of writing and reading, He writes:

> The lines inscribed on the page, [...] were the visible traces of dextrous movements of the hand. And the eye of the reader, roaming over the page like a hunter on the trail, would follow these traces as it would have followed the trajectories of the hand that made them.[34]

Lines, in this way of thinking, are not only the unfurling of matter along trajectories. A line, in the sense of a trajectoried entity, a living thing, a relationship, or a coherent idea, only appears for something or someone who stories it into being. It is written both in its unfurling, and by whom or whatever discerns the coherence and duration of it, its semiotic material narrative form as a directional story. It has the kind of both-ways-becoming discussed in Chapter 3. The storyteller or reader is not themselves a fixed point, they are also an unfurling line, as Ingold writes: "observations are taken not from points at all, but along paths of movement".[35]

This account integrates well with an understanding of materiality as semiotic, as being written and read with meaning by multiple writers and interpreters whose roles as such are continually interchanging. Knotting twists threads of life stories together and from within emerge relations, ethics, obligations towards mutual ongoing. What emerges as a thread, as a line, is already a matter of having the semiotic framework to read it. If lives and durations of being are conceptualised as becoming along lines, lines of desire, lines of flight, lines of unfolding gambles upon a chancy run at ongoingness, already these lines are thick, knotty with other lines inside them, finally knotting together more thickly into things like mangroves, cities, and forests.

Knots of obligation

Obligated orientations and knots of embodied time

In Chapter 3, I introduced a discussion of semiotic material obligation. Some thinkers have adopted a visualisation of these obligations as tangled or knotted lines to explain ethical considerations from a relational ontology. To understand its potential further, I will now discuss several of the concepts developed to explore this way of thinking.

A conceptualisation of obligated entangled lines is developed within the work of two writers both working within the context of Indigenous Australian ways of living with land. Rose, in her work on knots of embodied ethical time, and Povinelli in her thinking about obligated orientations, develop their discussions of a world that is knotted or entangled to think through ethical implications. Both provide accounts of ethics that relate to entanglement and commitment within the thick story lines of country.[36] Rose writes of the world of her Indigenous teachers: "the living world is made up of differentiation, pattern and connection. Everything is connected to some things and not to others, but everything is connected and nothing is left stranded".[37]

Povinelli introduces her readers to Tjipel, a mangrove estuary in Northern Australia.[38] Elders from the Karrabing group (a group identified around a project), taught Povinelli that Tjipel is a girl on the cusp of womanhood, who dressed and hunted as a boy. The estuary is the endplace of Tjipel's long and eventful journey across country. After a fight with an old man, Tjipel's body parts were strewn around that place, her spear became the fringing reef, her legs the streams running into the bay, all the parts of the land there are parts of Tjipel, who is between so many things – woman and girl, male and female, dead and alive, living and non-living, sea and river, water and land, the past and the present, human and non-human, one and many. As an estuary ringed with mangroves, Tjipel's betweenness is apparent upon the country. Povinelli notices that Tjipel has no skin – like other mangrove estuaries, any borders are continually crossed by so many of the community members, the sharks, the birds, the tides, for example. Karrabing people do not think Tjipel by her edges, as though she were inside a wall or skin. Rather, she is what Povinelli describes as the "thishereness" produced by the orientations of numerous plants, animals, particles, and movements towards each other. Tjipel is, Povinelli explains, "an assertion about a set of the obligated orientations without an enclosing skin".[39] Many beings;

water, mud, trees, fish, crabs, mosquitoes, entangle their journeys about each other such that Tjipel is able to become and continue. This "focus of habituated attention" in turn affects things, requiring certain engagements. Fish will swim following the path of her channel, mud will settle among her roots. Tjipel "exerts a habituating force"[40] by requiring that to participate within her, entities must maintain their attention, their orientation towards the others within her.

In Povinelli's account, the Karrabing are aware of being included within the obligated membership of Tjipel. As they do for all the sites of meaning in their country, Povinelli explains, they realise a need to pay attention, to read Tjipel's signs and respond to her needs, to care for her because they are obligated participants within her active interminglings. She is a host that provides plenty for them, but they are aware that she can withdraw her hospitality, turn her back, by becoming unproductive or by disappearing. If she withdraws it is because of her skinless vulnerability, because her signs have not been attended to by those obliged to her. Tjipel is made of entangled lines, and she also has her own journey and entanglements, and these also matter; they contribute onwards towards writing the country. Tjipel's journey across the country is affected by, and affects other lines through encounters along the way, meaning that paying attention to other obligations across country may affect Tjipel, because her story intersects with others.

"Knots of embodied time" is the lush expression Rose coins to describe the way individuals, species, and communities are better thought of as the obligated intertwinings of multiple temporalities.[41] The mosquito is the knot formed as intergenerational gifts such as genes, bacteria, and practices provided by ancestors and all those who nourished them, come together with various other temporalities, such as those of the plants, animals, and water that help to support her life; "life depends both on the sequential processes of generational time/gift and on the synchronous processes of multispecies nourishment. These processes and patterns intersect to form dense knots of embodied time".[42] Rose follows the flying-fox in its relationship with the eucalyptus blossom whose nectar it drinks, finding the unfurling futures of the trees and the animals storied onwards together across generations. "Within this wider world of multispecies knots," she writes,

> ethics may be understood as an interface—a site of encounter and nourishment. Thus the encounter between the flower of the tree and the tongue-and-fur of the flying-fox forms an interface where the desire for one's own life is shaped to the desire that others have for their own lives.[43]

Ethical responsiveness can be considered within the frame of such synchronous alignment of desires, in respect to the question of how well actions are contributing towards these life desires. Ethical responsiveness also, she argues, needs to consider the sequential, the responsibility to the gifts of intergenerational stories. Here Rose agrees with James Hatley's assessment of the ethical nature of the intergenerational gifts of heredity that story across individuals to connect them as a race or a species. Individuals are already in themselves a material responsibility within that story because they receive that hereditary story, materially, as a gift. Responsibility is to a particular articulation of time – an articulation of time within which the ethical terms of this particular intergenerational story hold: "the narrative breathed across generations arrives unasked for and carries an obligation".[44] Living beings live and die not as individual discrete existences, but *as responsibilities* within an intergenerational narrative.[45]

By this intergenerational, sequential-time-oriented account, ethicality is relative to what a particular thread or story of life has come to set up as its enterprise. Within that narrative, that which achieves that enterprise counts as ethical. But it seems important to remain critical of arguments that align ethics with an evaluation of success – whether success is measured against ecological or economic or aesthetic or any other evaluative framework. The synchronous-time account of ethicality as always responsive at the interfaces of affect provides a counter-narrative that can be understood as providing a continual critical re-evaluation of those intergenerational sequential narratives, in terms of their relevance or appropriateness in light of ethicality within the momentary interaction. The sequential narratives, in turn, provide a similar counter-narrative set of semiotic criteria for the critical evaluations of the synchronous inter-affective responses. Together the two temporalities provide directionality that is meaningful because of its continual self-reflexive criticality. In this way Rose's account of multispecies knots of embodied time suggests a model that has the potential to describe parameters within which a non-anthropocentric relational semiotic materiality critically assesses its ethicality, without recourse to "higher level" human judgement.

This process, with its each-way-cross-checking, is familiar. It has the same criss-crossed folding form that was described in "Strange Reflexivities" with Merleau-Ponty's account of perceptual awareness, Deleuze's account of immanent becoming, and the oyster's continual conversation with the tides. In each case, a developing material semiotic narrative (a sensory capacity, an oyster, or an inter-species symbiotic relationship) folds back into synchronous inter-responsive encounters offering criteria for the choices being taken within them (how oyster flesh responds to this particular fluid encounter, which blossom this flying-fox decides to visit, whether this mosquito bites this wallaby today) while these inter-responsive encounters fold back into the developing narrative, offering critiques, revisions, and amendments. Rose's work suggests an ecological ethicality located within the in-between of this criss-crossing; that ecological ethicality comprises this continual both-ways-cross-checking that reviews, refines, and changes the criteria for both the conduct of interactions in the here and now, and the longer-durational unfurling narrative lines of meaning.

Traps and reciprocal captures

Alberto Corsín Jiménez describes knots of obligation occurring as human societies develop ways of living that interface with their environments.[46] In a discussion that pays tribute to von Uexküll, he describes this as a "double captivity" imagined as a kind of semiotic spiderweb. A spiderweb interfaces the worlds of the spider and of its prey. It is meaningful and entangled in a discourse of attunement in two directions, speaking out into the world of the fly whose habits and history it has come to materially respond to, and speaking in towards the needs and capacities of the spider, whose physiology, desire, and behaviours are written into its materiality. Further, its doubleness binds these two worlds; this is what makes it a trap. Corsìn Jiménez proposes that between a group of humans and their environment, through continuous examinations and experiments in narratives of description, human attempts to describe and know spin traps that compose the relationship into meaningful tropes: predator and prey, host and guest, territory and community. These become entrapments, binding the people and the environment into their patterns of meaning. Central to entrapment, according to Corsìn Jiménez, is a recursive move, a flip. The trap demonstrates, opens, makes real, the "in-between" of two worlds and in doing so twists the two to face each other, each one's own patterns of being becoming attuned with the other's. Corsìn Jiménez writes of human cultures, but non-human cultures would presumably be just as sticky.

In a similar fashion, Isabelle Stengers considers obligation in her discussions of ecologies of practices, constraint, and reciprocal capture. Practices, she finds, get figured out on the run, finding themselves already obligated by the pressing demands to respond to whatever they are constrained and enabled by. What works is always a co-production which sees identities co-created within obligations-between that capture each party, meaning that capture is reciprocal. Further, "wherever there is reciprocal capture value is created" as obligations enable new possibilities, just as they constrain others.[47] Constraints are obligations that arise with the creation of value within the becoming-obligated of participating practices. "Obligation here refers to what obligates practitioners to what is 'required of a phenomenon' for it to be addressed as a focus for a particular practice", is Maria Puig de la Bellacasa's summary.[48] Stengers argues against the tendency to substitute a set identity in the place of entwined obligations.[49] Always things are obligated within practices and reciprocal captures, and these are the sources of any kind of value. "You think when forced or obliged to think. You do not think without a 'cause' ", she argues.[50] Logic itself she regards as an adapted practice, adopted through a need to respond to an obligation. It is not a "matter of generality" outside of obligated entanglements: "what you are responsible for is paying attention as best you can, to be as discerning, as discriminating as you can about the particular situation".[51] Despret and Meuret illustrate Stengers' ideas through their example of the flock. Ethics in the flock is about "ways of being obligated by those to whom you give the power to obligate you. What affects you, and whom does your way of living, your manner of being, affect in turn?"[52] Here obligation is a tender, lived sensitivity that is productive of emergent value. From obligation a collective flock memory emerges of how to live as flock and "by the concrete memory of the mouths, the eyes, the guts, the bodies, the legs, and the feet, the flock multiplies the ways lands, paths, bushes, springs, and rocks exist".[53] What should concern us, Stengers argues, is how to be more sensitively obligated because this is how value increases.

A semiotic material ethics of obligation

These four accounts of obligation, taken together, illustrate a model of ethics as sensitivity to the obligations entangled within a semiotic materiality. Corsìn Jiménez describes the stickiness of double captivities between cultures and natures, Povinelli illustrates the entanglements of storying within obligated orientations, and Stengers highlights the importance of context, openness, and acute attentiveness. Puig de la Bellacasa interprets the obligations of such entanglements as always immanent: "The ways in which these entanglements affirm what is of 'value' is an immanent one, … one relational practice does not equate another".[54] Every knot obligates its participants within an ethics that is made inside the tangle. Rose's acknowledgement of the ethics of sequential time, however, reminds us that material narratives are being written at numerous temporal and spatial scales, and that these are what are entangling. The synchronous and sequential cross-referencing discussed by Rose suggests processes of ethical evaluation that situate obligations and value beyond a singular immanence. Processes of evaluation occur as a both-ways-cross-checking and interfoldedness between material semiotic scales of narrative becoming (not limited to those two particular temporalities that Rose discusses). Some cross-referencing is involved across scales for there to be criteria for evaluation. But here these scales are interfolded, not rigid and hierarchically layered with those at a higher level responsible for those below; they continually shimmer with both-ways inter-responsiveness across narrative scales.

My body is deeply entangled in relations with a mosquito, a virus and a wallaby, it is drawn into various actions and decisions because of these obligations. My immune system is obliged to engage, my hearing is obliged to become sensitive to mosquito whine, I am obliged to itch; the relationship has defined its own criteria of duty. But the obligations of my immunity are responsive and reflexively evaluative, not rule-like. They are guided by the patterns and complexities of the past, but critically reassessed by the particularities of inter-responsive affect at every encounter.

Obligation is never free of trouble; it tugs against freedom and involves constraint. The pleasures of flying-foxes and honey blossoms, or the wasps taking pleasures within their orchids, suggest that obligated co-becomings can point the way towards mutually nourishing relations, but equally relevant are all those who suffer within repressive or debilitating relations of obligation; the coerced, the domesticated, the infantilised, the colonised, the infected, the autoimmune. Becoming committed to relations of mutual obligation with other entities offers a guide towards ethical engagement that is always troubled and so requires continual critique. It requires that obligations be worried over, that they become matters of dedicated concern, and that they are checked against the obligations of numerous narratives at various scales. Such an account of ethics cannot provide an untethered ethical guidance, but suggests where attention needs to be paid, to the interfaces of co-becoming obligations, and to the knotted relational stories these pan out across.[55] Some knots are thicker and more supportive of many more obligated narratives than others. The mangroves of the Cooks River, for example, entangle and support many more narratives than the concrete-edged channel that has replaced them along much of that river, a discussion which will be expanded through the final section of this book.

Relations of mutual attentiveness within reciprocal co-becoming are not reliable, nor are they the effortless outcomes of genetic and evolutionary algorithms. They demand both physical energy and semiotic work. It is all too obvious that many of the ecological obligations established over generations of inter-responsive co-becomings are no longer receiving the attentions adequate for their continuance. There is no tree hollow where it is needed, no mudflats to rest and feed after the long migratory flight, no insect swarm to pollinate the blossom, no water in the land for roots to find. Obligations are often burdensome, tiring, or painful. Where an option becomes available for avoiding the unpleasant obligation while still receiving the benefit, this could perhaps be rewritten as a narrative of a gift to be seized or a chance to be exploited (frequent tropes in human/nature narratives). Some human semiotic accounts found a way to stop paying inherited debts, we disobliged ourselves, stole and enslaved to meet our needs, and made a run at an imaginary freedom.

Relations are stories in a material semiotics

Relations/stories

Within a materiality understood as semiotic, as full of meanings being made and interpreted, relations, I suggest, are obligations written into narratives. They are stories situated meaning-makers write – with bodies, inside bodies, among bodies, including bodies that are vegetative bacterial, watery, vaporous, social, or scriptural – to construe as meaningful the processes of intra-responsive co-becoming and becoming complexly obligated, stirred up by regularities and disturbances. A relation is its material telling, its storying-on.[56] I am using the words narrative and story loosely, not implying something tightly defined with a beginning, middle, and end, or a certain kind of plot. I am thinking of a narrative as the signs, syntax, semiotics, and forms of memory, that develop to sustain particular situated meanings – whether in a cell by viral RNA or within data by an epidemiologist. Timothy Morton suggests "poem" might be a more apt descriptor,[57] and I would not disagree, though I think that within a semiotic material context, the distinction between poetry and story becomes moot. I also clarify that storying as discussed here is not referring to a part of relations that is attached to human cultural work, while the material part sits outside, only storied into meaningfulness by a human. That would not reflect a semiotic material ontology.

Through my reading of mosquito, virus, and mangrove research, I have found many richly woven storyings of particular events, phenomena, or places. Some are scientific accounts about the awe inspiringly complex strategies of viruses as they and their hosts reconstruct themselves and each other through changing circumstance,[58] some are about the world-changing roles of mosquitoes, or mosquitoes and viruses together, interwoven among many other actants, human and more-than-human, within a particular situation,[59] and some are tales of the tidings of mangroves as they insinuate or bash their ways into human, political, and ecological complexes.[60] All these accounts are attentive, detailed investigations of interactions between numerous actants that acknowledge collective and non-human action and demonstrate how relational narratives become written and knotted. I will recount two of these briefly, as though to suggest their interweaving into ever more layered narratives of performative reality. These multistranded accounts and the suggestion that they in turn can be interwoven acknowledge Haraway's insights concerning the methodological strength of understanding knowledge as situated and strengthened from being interwoven.

Writing of the possibilities for the mosquito to speak, Timothy Mitchell weaves together many of the narrative threads and discursive practices involved in the devastating malaria outbreak in Egypt in 1942.[61] The account folds in dam engineering theories and practices, the changing flows of the Nile caused by damming, the influx of new synthetic chemical fertilisers and the political intrigue of those involved within the burgeoning fertiliser industry, the bullish ambitions for profit of superwealthy capitalists and the institutional arrangements of state that supported their ambitions, war and its dislocations, trade impacts and demands upon the country's resources, the impact of industrialised agriculture on poor farming populations, and importantly, mosquito behaviours and travel strategies[62] and the adaptability of *Plasmodium falciparum*, the principal malaria parasite, in finding opportunity within the chaos of war, famine, and greed. "Dams, blood-borne parasites, synthetic chemicals, mechanized war, and man-made famine coincided and interacted", Mitchell writes.[63] He observes that only by a research methodology that holds open the issue of power and agency, and allows their form and impact to be a question, answerable through acknowledging and attending to as many "hybrid agencies, connections, interactions, and forms of violence" as possible,[64] can an account emerge that can escape the well-dug ruts of disciplinary paradigms, which redirect evidence towards narrow human-oriented modes of evaluation. Further, he suggests decision-making that ignored complex interactivity was complicit in "producing the effect of neatly separate realms of reason and the real world, ideas and their objects, the human and the nonhuman" that ultimately lay behind disasters such as the Egyptian malaria outbreak, and numerous other human and environmental disasters of the twentieth century.[65] Although his analysis goes beyond this, Mitchell's scholarship is one example of many recent histories detailing how human societies have been shaped by pathogens. Whittaker summarises their general findings that

> pathogens that cause human disease have greatly affected the way humans live for centuries. Many infectious diseases have been significant enough to affect how and where we live, our economies, our cultures and daily habits. And many of these effects continue long after the diseases have been eliminated.[66]

Heather Goodall, Allison Cadzow, and Denis Byrne's performative account of the ways mangroves and saltmarsh have participated in developing particular ecological and fluvial environments along the banks of the Georges River also weaves together more-than-human stories, in this case, to build a complex history of these riverbanks.[67] They explain that the two ecological communities interacted in

> see-sawing movements up and down the bank, depending on minute sea-level changes to alter the salinity of the soil, or on the numbers of grazers like swamp wallaby at any one time, which was in turn dependent on the number of human hunters, and so on.[68]

Saltmarsh and mangrove stories entangle together, and with those of hunters and swamp wallabies. As Sydney's southern suburbs and industrial areas grew through the late nineteenth and twentieth centuries, the stories of urban settlement joined and complicated these patterns. As silt and nutrients from land clearing, agriculture and industry increased in the Georges River valley, the river's mangroves responded by expanding their reach. At the same time, much of the river's saltmarsh was becoming degraded through pressures from development, so in some areas, mangroves expanded into the degraded saltmarsh zones. Popular opinion described this expansion as invasive and characterised the mangroves as aggressive towards the ever-rarer saltmarsh. This combined with a long-standing association of mangrove swamps with illness – a legacy of superseded theories that linked the spread of diseases with miasma – meaning that through much of the twentieth century the mangroves of this region were widely regarded among those humans responsible for the river's management as not only "an enemy of health, hygiene and modernity",[69] but also ecologically malevolent. Mangroves thus became the "bad guy" relative to saltmarsh, in the narratives of those whose agendas required a justification for their removal.

These two carefully researched studies both demonstrate how obligations to particular values formed within the entanglements of more-than-human stories. In each case the various narrative threads were diverse, involving highly divergent forms of semiotic material evaluations across differing temporal and spatial scales.

Vector

Aedes vigilax is a vector for the RRV virus. Many relations involve vectors to perform a boundary crossing process. A vector opens a pathway and delivers the guest, the parasite, or the gift. It provides a disruptor and a connector to another story.[70] A vector offers a narrative mechanism for experimenting across the patterns that have established within a relationship or entity. It dramatically expands a story, to open it to a new scale of engagement, to make a major twist in direction.

As a vector, our mosquito is not simple. She is a complex body, an assemblage made of multiple inter-relating components. During feeding, when it is ingested with the blood meal, the virus enters the mosquito's gut. In order to survive there and increase in number, the virus must infect cells and replicate.[71] Various environmental factors in the midgut affect whether this happens successfully. For example, the presence of a bacteria called Wolbachia within the mosquito's gut diminishes the successful reproduction of a number of viruses similar to RRV, including dengue, Zika, and Chikungunya[72] (research for RRV specifically had not been reported at the time of writing[73]). Because of these effects, Wolbachia infected mosquitoes are being released into dengue-endemic zones, including sites in Queensland in Australia, where it was not present within the native mosquitoes.[74] While it reduces reproduction of these viruses, the presence of Wolbachia in a mosquito microbiome favours the transmission of some Plasmodium species which are associated with malaria in birds and rodents.[75] After the midgut, the virus must cross the basal lamina, the barrier between the gut and the rest of the body. Research has found that mosquitoes which were poorly nourished as larvae may have a thinner barrier here, making it easier for the virus to cross.[76] Well-nourished mosquitoes have thicker gut walls, and fewer viruses can pass across, although the presence of a blood meal softens the barrier making it easier to cross. Once through this barrier the virus infects cells through other parts of the mosquito's body, where it faces the mosquito's immune system. Passing through another barrier into the salivary glands is the next challenge for the virus. Once in the salivary glands, the virus replicates quickly and waits. The next chapter will follow the virus as it moves from the mosquito saliva across my skin during the bite.

Notes

1 Val Plumwood, *Feminism and the Mastery of Nature* (London and New York: Routledge, 1993).; Donna Haraway, "Situated Knowledges: The Science Question in Feminism and the Privilege of Partial Perspective," *Feminist Studies* (1988), https://dx.doi.org/10.2307/3178066
2 Carol Gilligan, *In a Different Voice: Psychological Theory and Women's Development* (Cambridge, Massachussetts: Harvard University Press, 1982).
3 For example, Elizabeth Grosz, *Volatile Bodies: Towards a Corporeal Feminism* (Bloomington Indiana: Indiana University Press, 1994).
4 For example, Susan Griffin, *Woman and Nature: The Roaring inside Her* (New York: Harper & Row, 1978); Carolyn Merchant, *The Death of Nature: Women, Ecology, and the Scientific Revolution* (San Francisco: Harper & Row, 1980).
5 An example of this approach is Mustafa Emirbayer, "Manifesto for a Relational Sociology," *American Journal of Sociology* 103, no. 2 (1997), www.jstor.org/stable/10.1086/231209
6 François Dépelteau, "Relational Thinking in Sociology: Relevance, Concurrence and Dissonance," in *The Palgrave Handbook of Relational Sociology*, ed. François Dépelteau (Cham: Springer International Publishing AG, 2018), 9. The definition is cited in C. Watts, *Relational Archeologies: Humans, Animals, Things*, (London: Routledge. 2013).
7 I recognise that many versions of relational ontologies would not make this connection.
8 Bryan Bannon, "Developing Val Plumwood's Dialogical Ethical Ontology and Its Consequences for a Place-Based Ethic," *Ethics & the Environment* 14, no. 2 (2009): 48.
9 Bannon, "Developing Val Plumwood's Dialogical Ethical Ontology", 50.
10 A. E. Wright, *Mosquito Life Cycles, Ecology and Behaviour* (Ballina: Ballina Shire Council, 2002).
11 Cameron Webb, personal communication, 5 August 2018.
12 Cameron Webb, "Ross River Virus in Melbourne, How Did That Happen?", *Mosquito Research and Management* (2017), accessed 4 July 2019, https://cameronwebb.wordpress.com/2017/02/10/ross-river-virus-in-melbourne-how-did-that-happen/
13 Suzi B. Claflin and Cameron E. Webb, "Ross River Virus: Many Vectors and Unusual Hosts Make for an Unpredictable Pathogen," *PLoS Pathogens* 11, no. 9 (Sep 2015), https://dx.doi.org/10.1371/journal.ppat.1005070; David Harley, Adrian Sleigh, and Scott Ritchie, "Ross River Virus Transmission, Infection, and Disease: A Cross-Disciplinary Review", *Clinical Microbiology Reviews* 14, no. 4 (2001), https://dx.doi.org/10.1128/CMR.14.4.909-932.2001
14 Wright, *Mosquito*.
15 Richard C. Russell, "Ross River Virus: Ecology and Distribution," *Annual Review of Entomology* 47, no. 1 (2002), https://dx.doi.org/10.1146/annurev.ento.47.091201.145100; Eloise B. Stephenson et al., "The Non-Human Reservoirs of Ross River Virus: A Systematic Review of the Evidence," *Parasites and Vectors* 11, no. 1 (2018), https://dx.doi.org/10.1186/s13071-018-2733-8
16 Charles. M. El-Hage, M. J. McCluskey, and Joseph K. Azuolas, "Disease Suspected to Be Caused by Ross River Virus Infection of Horses," *Australian Veterinary*

Journal 86, no. 9 (2008), https://dx.doi.org/10.1111/j.1751-0813.2008.00339; Courtney Wilson, "Horses Succumbing to Ross River Virus in Queensland, Affecting Equestrian Events and Recreational Riders," *ABC News, Australian Broadcasting Commission*, 2015, accessed 4 July 2019, www.abc.net.au/news/2015-05-16/increasing-number-of-horses-succumbing-to-ross-river-virus/6472620

17 Jade Redfern, "Differential Profiles of Mammalian and Mosquito Derived Ross River Virus" (University of Canberra, 2015).

18 Harvey Lodish, Arnold Berk, and S. Lawrence Zipursky, "Viruses: Structure, Function, and Uses," in *Molecular Cell Biology*: 191-207, ed. Harvey Lodish et al. (New York: W. H. Freeman, 2000).

19 "Are you a multispecies entity, made up of some human bits and some microbial bits – or are you just the human bits, with an admittedly fuzzy boundary between yourself and your tiny companions?" asks Derek Skillings in, "I, Holobiont. Are You and Your Microbes a Community or a Single Entity?", *Aeon Ideas*, 2018, https://aeon.co/ideas/i-holobiont-are-you-and-your-microbes-a-community-or-a-single-entity. The question can equally be asked of a mosquito.

20 On the mosquito microbiota see Eva Novakova et al., "Mosquito Microbiome Dynamics, a Background for Prevalence and Seasonality of West Nile Virus," *Frontiers in Microbiology* 8, no. 526 (2017), https://dx.doi.org/10.3389/fmicb.2017.00526

21 Throughout this chapter, "affect" is used as it was described in Chapter 2, and much influenced by Elizabeth Grosz, as a general term for any impact or influence. I cannot add "of one thing upon another" because it is not always the case that things exist before affect.

22 Gilles Deleuze and Felix Guattari. *A Thousand Plateaus: Capitalism and Schizophrenia*, trans. Brian Massumi (Minneapolis and London: University of Minnesota Press, 1987), 167.

23 Deleuze and Guattari, *Plateaus*, 215.

24 Ibid., 229.

25 Ibid., 272.

26 Ibid., 161.

27 There appears to be a high degree of redundancy built into a mosquito's capacity to smell a potential host. Mosquito smell receptors are able to respond to many of the chemical odours humans emit, so if one is absent, others can be used, and one mix may be used to discriminate a favoured species from another. See Helen H. Shen, "Inner Workings: How Do Mosquitoes Smell Us? The Answers Could Help Eradicate Disease," *Proceedings of the National Academy of Sciences of the United States of America* 114, no. 9 (Feb 2017), https://dx.doi.org/10.1073/pnas.1701738114

28

> from 10 to 50 meters away, a mosquito smells a host's CO_2 plume. As it flies closer – to within 5 to 15 meters – it begins to see the host. Then, guided by visual cues that draw it even closer, the mosquito can sense the host's body heat. This occurs at a distance of less than a meter,

in Jessica Stoller-Cohen, *Mosquitoes Use Smell to See Their Hosts*, (Caltec, 2015), www.caltech.edu/about/news/mosquitoes-use-smell-see-their-hosts-47338

29 Harley, Sleigh, and Ritchie, "Ross River", 118.

30 Diver, cited in Phil McManus, "Mangrove Battlelines: Culture/Nature and Ecological Restoration," *Australian Geographer* 37, no. 1 (2010): 69, https://dx.doi.org/10.1080/00049180500511970
31 Norman Duke and Klaus Schmitt. "Mangroves: Unusual Forests at the Seas Edge," in *Tropical Forestry Handbook*, ed. Laslo Pancel and Michael Köhl (Berlin: Springer-Verlag, 2015): 1709–1710.
32 Tim Ingold, "Bindings against Boundaries: Entanglements of Life in an Open World," *Environment and Planning A: Economy and Space* 40, no. 8 (2008): 1796, https://dx.doi.org/10.1068/a40156
33 Tim Ingold, *The Life of Lines* (Oxon and New York: Routledge, 2015), 16.
34 Ingold, *The Life of Lines*, 27.
35 Ingold, "Bindings", 1800, note 3.
36 Country is an important concept in many Australian Indigenous philosophies which is being adopted into a wide range of Australian social-ecological discourses. Sarah Wright describes the concept as "an integrated, more-than-human presence that incorporates land, animals and people, but also non-human beings such as tides, waters, winds, insects, rocks, plants, languages, emotions, songs and ancestors", in "More-Than-Human, Emergent Belongings," *Progress in Human Geography* 39, no. 4 (2014): 403, https://dx.doi.org/10.1177/0309132514537132. Wright outlines an emergent conceptualisation of belonging that is based upon the obligations generated within emergent co-becomings: "A person belongs to Country and to an ant and a tree the same way that Country, the ant, the tree and the song belong to a person, because there is no separating out their identities", in "More-Than-Human", 403.
37 Deborah Bird Rose, "Judas Work," *Environmental Philosophy* 5, no. 2 (2008): 58, https://dx.doi.org/10.5840/envirophil200741/217
38 Elizabeth A. Povinelli, *Geontologies: A Requiem to Late Liberalism* (Durham and London: Duke University Press, 2016), chapter 4, Kindle.
39 Povinelli, *Geontologies*, chapter 4, paragraph 18.
40 Ibid.
41 Donna Haraway also frequently writes of relations of co-becoming as knots in lived trajectories. For example: "the infolding of others to one another is what makes up the knots we call beings", she writes in *When Species Meet*. (Minneapolis: University of Minnesota Press, 2008), 250, illustrating how she regards knotting and folding as complementary.
42 Deborah Bird Rose, "Multispecies Knots of Ethical Time," *Environmental Philosophy* 9, no. 1 (2012): 129.
43 Rose, "Multispecies", 136.
44 Ibid., 130.
45 Ibid., 135.
46 Alberto Corsín Jiménez, "Spiderweb Anthropologies: Ecolocies, Infrastructures, Entanglements", in *A World of Many Worlds*, ed. Marisol De La Cadena and Mario Blaser (Durham: Duke University Press, 2018).
47 Isabelle Stengers, *Cosmopolitics* trans. Robert Bononno, vol. 1, 2 vols. (Minneapolis: University of Minnesota Press, 2010); Isabelle Stengers, "Introductory Notes on an Ecology of Practices," *Cultural Studies Review* 11, no. 1 (2005): 36.

48 María Puig de la Bellacasa. *Matters of Care: Speculative Ethics in More Than Human Worlds* (Minneapolis and London: University of Minnesota Press, 2017), 153.
49 Stengers, *Cosmopolitics*, 211.
50 Stengers, "Introductory", 191.
51 Ibid., 188.
52 Vinciane Despret and Michel Meuret, "Cosmoecological Sheep and the Arts of Living on a Damaged Planet," *Environmental Humanities* 1, no. 8 (2016): 25–36.
53 Despret and Meuret, "Sheep", 33.
54 Puig de la Bellacasa, *Matters*, 52–53.
55 Despret and Meuret, "Sheep", 35; Deborah Bird Rose, "Val Plumwood's Philosophical Animism: Attentive Interactions in the Sentient World," *Environmental Humanities* 3, no. 1 (2013): 104, https://dx.doi.org/10.1215/22011919-3611248
56 Serenella Iovino and Serpil Oppermann describe the "stories of matter" as characterised by "a dynamic process of material expressions seen in bodies, things, and phenomena coemerging from … networks of intra-acting forces and entities", in "Stories Come to Matter," in *Material Ecocriticism*, ed. Serenella Iovino and Serpil Oppermann (Bloomington and Indianapolis: Indiana University Press, 2014), 7.
57 Timothy Morton, "The Liminal Space between Things: Epiphany and the Physical," in *Material Ecocriticism*, ed. Serenella Iovinno and Serpil Oppermann (Bloomington and Indianapolis: Indiana University Press, 2014), 269.
58 Laurence Briant, Philippe Desprès, Valérie Choumet, and Dorothée Missé, "Role of Skin Immune Cells on the Host Susceptibility to Mosquito-Borne Viruses," *Virology* 464–65 (2014), https://doi.org/10.1016/j.virol.2014.06.023; Merry Youle, *Thinking Like a Phage: The Genius of the Viruses That Infect Bacteria and Archaea* (San Diego: Wholon, 2017), Kindle.
59 Ann H. Kelly and Javier Lezaun, "Urban Mosquitoes, Situational Publics, and the Pursuit of Interspecies Separation in Dar Es Salaam," *American Ethnologist* 41, no. 2 (May 2014), https://dx.doi.org/10.1111/amet.12081; John McNeil, *Mosquito Empires: Ecology and War in the Greater Caribbean 1620–1914* (Cambridge: Cambridge University Press, 2010); Timothy Mitchell, *Rule of Experts: Egypt, Techno-Politics, Modernity* (Berkeley and Los Angeles: University of California Press, 2002); Emily O'Gorman, "Imagined Ecologies: A More-Than-Human History of Malaria in the Murrumbidgee Irrigation Area, New South Wales, Australia, 1919–45," *Environmental History* 22, no. 3 (2017), https://dx.doi.org/10.1093/envhis/emx056
60 Heather Goodall, Allison Cadzow, and Denis Byrne. "Mangroves, Garbage, Fishing: Bringing Everyday Ecology to Sydney's Industrial Georges River," *Transforming Cultures eJournal* 5, no. 1 (2010): 52–83; Laura A. Ogden, *Swamplife: People, Gators, and Mangroves Entangled in the Everglades* (Minneapolis: University of Minnesota Press, 2011), Kindle.
61 Mitchell, *Rule of Experts*, chapter 1. Mitchell cites the figures of three-quarters of a million people contracting malaria during this outbreak, and between one and two hundred thousand dying from the disease. Most affected were the poor workers on the sugar estates; at 20.
62 The *Anopheles gambiae* mosquitoes are thought to have arrived via airplanes, and spread to new areas via boat, train and road transport; at ibid., 23.

63 Ibid., 22.
64 Ibid., 53.
65 Ibid., 52.
66 Maxine Whittaker, "How Infectious Diseases Have Shaped Our Culture, Habits and Language", *The Conversation*, 2017, theconversation.com/how-infectious-diseases-have-shaped-our-culture-habits-and-language-75061
67 Goodall, Cadzow, and Byrne, "Mangroves, Garbage, Fishing."
68 Ibid., 77.
69 Ibid.
70 A fuller discussion of the concepts of host, guest, gift, and parasite occurs in Part 3 of the book.
71 Claudia Rückert and Gregory D. Ebel, "How Do Virus–Mosquito Interactions Lead to Viral Emergence?", *Trends in Parasitology* 34, no. 4 (2018), https://dx.doi.org/10.1016/j.pt.2017.12.004
72 Ryan Kerney, Zakiya Whatley, Sarah Rivera, and David Hewitt, "The Prospects of Artificial Endosymbioses", *American Scientist*, 2017.
73 Cameron Webb, personal communication, 20 August 2018.
74 Queensland Government, "Dengue Safe Project Information", in *The State of Queensland* (Queensland Health, 2019), accessed 25 July 2019, www.health.qld.gov.au/townsville/services/tropical-public-health/ingham-dengue-safe/project
75 Grant L. Hughes, Ana Rivero, and Jason L. Rasgon, "Wolbachia Can Enhance Plasmodium Infection in Mosquitoes: Implications for Malaria Control?", *PLoS Pathogens* 10, no. 9 (Sep 2014), https://dx.doi.org/10.1371/journal.ppat.1004182; Novakova et al., "Mosquito Microbiome".
76 Rückert and Ebel, "How Do", 313.

References

Bannon, Bryan. "Developing Val Plumwood's Dialogical Ethical Ontology and Its Consequences for a Place-Based Ethic." *Ethics & the Environment* 14, no. 2 (2009): 39–55.

Briant, Laurence, Philippe Desprès, Valérie Choumet, and Dorothée Missé. "Role of Skin Immune Cells on the Host Susceptibility to Mosquito-Borne Viruses." *Virology* 464–65 (2014): 26–32. https://doi.org/10.1016/j.virol.2014.06.023

Claflin, Suzi B. and Cameron E. Webb. "Ross River Virus: Many Vectors and Unusual Hosts Make for an Unpredictable Pathogen." *PLoS Pathogens* 11, no. 9 (Sep 2015): e1005070. https://dx.doi.org/10.1371/journal.ppat.1005070

Corsín Jiménez, Alberto. "Spiderweb Anthropologies: Ecologies, Infrastructures, Entanglements." In *A World of Many Worlds*, edited by Marisol De La Cadena and Mario Blaser, 53–82. Durham: Duke University Press, 2018.

Deleuze, Gilles and Felix Guattari. *A Thousand Plateaus: Capitalism and Schizophrenia*. Translated by Brian Massumi. Minneapolis and London: University of Minnesota Press, 1987. Originally published as *Mille Plateaux: Capitalism et Schizophréne 2*, Les Éditions de Minuit, 1980.

Dépelteau, François. "Relational Thinking in Sociology: Relevance, Concurrence and Dissonance." In *The Palgrave Handbook of Relational Sociology*, edited by François Dépelteau, 3–33. Cham: Springer International Publishing AG, 2018.

Despret, Vinciane and Michel Meuret. "Cosmoecological Sheep and the Arts of Living on a Damaged Planet." *Environmental Humanities* 1, no. 8 (2016): 25–36. https://dx.doi.org/10.1215/22011919-3527704

Duke, Norman and Klaus Schmitt. "Mangroves: Unusual Forests at the Seas Edge." In *Tropical Forestry Handbook*, edited by Laslo Pancel and Michael Köhl, 1698–1724. Berlin: Springer-Verlag, 2015.

El-Hage, Charles. M., M. J. McCluskey, and Joseph K. Azuolas. "Disease Suspected to Be Caused by Ross River Virus Infection of Horses." *Australian Veterinary Journal* 86, no. 9 (2008): 367–70. https://dx.doi.org/10.1111/j.1751-0813.2008.00339

Emirbayer, Mustafa. "Manifesto for a Relational Sociology." *American Journal of Sociology* 103, no. 2 (1997): 281–317. www.jstor.org/stable/10.1086/231209

Gilligan, Carol. *In a Different Voice: Psychological Theory and Women's Development*. Cambridge: Harvard University Press, 1982.

Goodall, Heather, Allison Cadzow, and Denis Byrne. "Mangroves, Garbage, Fishing: Bringing Everyday Ecology to Sydney's Industrial Georges River." *Transforming Cultures eJournal* 5, no. 1 (2010): 52–83. http://epress.lib.uts.edu.au/journals/TfC

Griffin, Susan. *Woman and Nature: The Roaring inside Her*. New York: Harper & Row, 1978.

Grosz, Elizabeth *Volatile Bodies: Towards a Corporeal Feminism*. Bloomington: Indiana University Press, 1994.

Haraway, Donna. *When Species Meet*. Minneapolis: University of Minnesota Press, 2008.

Harley, David, Adrian Sleigh, and Scott Ritchie. "Ross River Virus Transmission, Infection, and Disease: A Cross-Disciplinary Review." *Clinical Microbiology Reviews* 14, no. 4 (2001): 909–32. https://dx.doi.org/10.1128/CMR.14.4.909-932.2001

Hughes, Grant L., Ana Rivero, and Jason L. Rasgon. "Wolbachia Can Enhance Plasmodium Infection in Mosquitoes: Implications for Malaria Control?", *PLoS Pathogens* 10, no. 9 (Sep 2014): e1004182. https://dx.doi.org/10.1371/journal.ppat.1004182

Ingold, Tim. "Bindings against Boundaries: Entanglements of Life in an Open World." *Environment and Planning A: Economy and Space* 40, no. 8 (2008): 1796–810. https://dx.doi.org/10.1068/a40156

———. *The Life of Lines*. Oxon and New York: Routledge, 2015.

Iovino, Serenella and Serpil Oppermann. "Stories Come to Matter." In *Material Ecocriticism*, edited by Serenella Iovino and Serpil Oppermann, 1–18. Bloomington and Indianapolis: Indiana University Press, 2014.

Kelly, Ann H. and Javier Lezaun. "Urban Mosquitoes, Situational Publics, and the Pursuit of Interspecies Separation in Dar Es Salaam." *American Ethnologist* 41, no. 2 (May 2014): 368–83. https://dx.doi.org/10.1111/amet.12081

Kerney, Ryan, Zakiya Whatley, Sarah Rivera, and David Hewitt. "The Prospects of Artificial Endosymbioses." *American Scientist*. 105, no. 1 (2017): 36, 38–45.

Lodish, Harvey, Arnold Berk, and S. Lawrence Zipursky. "Viruses: Structure, Function, and Uses." In *Molecular Cell Biology*, edited by Harvey Lodish, Arnold Berk, S. Lawrence Zipursky, et al., 191–207. New York: W. H. Freeman, 2000.

Marr, Natalie, Mirjami Lantto, Maia Larsen, Kate Judith, Sage Brice, Jessica Phoenix, Catherine Oliver, Olivia Mason and Sarah Thomas. "Sharing the Field: Reflections of More-Than-Human Field/Work Encounters." *GeoHumanities* (2022). https// doi.org/10:1080/2373566X.2021.2016467

McManus, Phil. "Mangrove Battlelines: Culture/Nature and Ecological Restoration." *Australian Geographer* 37, no. 1 (2010): 57–71. https://dx.doi.org/10.1080/00049180500511970

McNeil, John. *Mosquito Empires: Ecology and War in the Greater Caribbean 1620–1914*. Cambridge: Cambridge University Press, 2010.

Merchant, Carolyn. *The Death of Nature: Women, Ecology, and the Scientific Revolution*. San Francisco: Harper & Row, 1980.

Mitchell, Timothy. *Rule of Experts: Egypt, Techno-Politics, Modernity*. Berkeley and Los Angeles: University of California Press, 2002.

Morton, Timothy. "The Liminal Space between Things: Epiphany and the Physical." In *Material Ecocriticism*, edited by Serenella Iovino and Serpil Oppermann, 269–279. Bloomington and Indianapolis: Indiana University Press, 2015.

Novakova, Eva, Douglas C. Woodhams, Sonia M. Rodríguez-Ruano, Robert M. Brucker, et al. "Mosquito Microbiome Dynamics, a Background for Prevalence and Seasonality of West Nile Virus." *Frontiers in Microbiology* 8, no. 526 (2017). https://dx.doi.org/10.3389/fmicb.2017.00526

Ogden, Laura A. *Swamplife: People, Gators, and Mangroves Entangled in the Everglades*. Minneapolis: University of Minnesota Press, 2011, Kindle.

O'Gorman, Emily. "Imagined Ecologies: A More-Than-Human History of Malaria in the Murrumbidgee Irrigation Area, New South Wales, Australia, 1919–45." *Environmental History* 22, no. 3 (2017): 486–514. https://dx.doi.org/10.1093/envhis/emx056

Plumwood, Val. *Feminism and the Mastery of Nature*. London and New York: Routledge, 1993, Kindle.

Povinelli, Elizabeth A. *Geontologies: A Requiem to Late Liberalism*. Durham and London: Duke University Press, 2016, Kindle.

Puig de la Bellacasa, María. *Matters of Care: Speculative Ethics in More Than Human Worlds*. Minneapolis and London: University of Minnesota Press, 2017, Kindle.

Queensland Government. "Dengue Safe Project Information." The State of Queensland (Queensland Health). 2019. Accessed 25 July 2019. www.health.qld.gov.au/townsville/services/tropical-public-health/ingham-dengue-safe/project

Redfern, Jade. "Differential Profiles of Mammalian and Mosquito Derived Ross River Virus." University of Canberra, 2015.

Rose, Deborah Bird. "Judas Work." *Environmental Philosophy* 5, no. 2 (2008). https://dx.doi.org/10.5840/envirophil200741/217

———. "Multispecies Knots of Ethical Time." *Environmental Philosophy* 9, no. 1 (2012): 127–40.

———. "Val Plumwood's Philosophical Animism: Attentive Interactions in the Sentient World." *Environmental Humanities* 3, no. 1 (2013). https://dx.doi.org/10.1215/22011919-3611248

Rückert, Claudia and Gregory D. Ebel. "How Do Virus–Mosquito Interactions Lead to Viral Emergence?" *Trends in Parasitology* 34, no. 4 (2018): 310–21. https://dx.doi.org/10.1016/j.pt.2017.12.004

Russell, Richard C. "Ross River Virus: Ecology and Distribution." *Annual Review of Entomology* 47, no. 1 (2002): 1–31. https://dx.doi.org/10.1146/annurev.ento.47.091201.145100

Shen, Helen H. "Inner Workings: How Do Mosquitoes Smell Us? The Answers Could Help Eradicate Disease." *Proceedings of the National Academy of Sciences of the United States of America* 114, no. 9 (Feb 28 2017): 2096–98. https://dx.doi.org/10.1073/pnas.1701738114

Skillings, Derek. "I, Holobiont. Are You and Your Microbes a Community or a Single Entity?" *Aeon Ideas*. 2018. https://aeon.co/ideas/i-holobiont-are-you-and-your-microbes-a-community-or-a-single-entity

Stengers, Isabelle. *Cosmopolitics*. Translated by Robert Bononno. Minneapolis: University of Minnesota Press, 2010.

———. "Introductory Notes on an Ecology of Practices." *Cultural Studies Review* 11, no. 1 (2005): 183–96.

Stephenson, Eloise B., Alison J. Peel, Simon A. Reid, Cassie C. Jansen, et al. "The Non-Human Reservoirs of Ross River Virus: A Systematic Review of the Evidence." *Parasites and Vectors* 11, no. 1 (2018): 188. https://dx.doi.org/10.1186/s13071-018-2733-8

Stoller-Cohen, Jessica. *Mosquitoes Use Smell to See Their Hosts*. Caltec. 2015. www.caltech.edu/about/news/mosquitoes-use-smell-see-their-hosts-47338

Webb, Cameron. "Ross River Virus in Melbourne, How Did That Happen?" *Mosquito Research and Management*, 2017. Accessed 4 July 2019. https://cameronwebb.wordpress.com/2017/02/10/ross-river-virus-in-melbourne-how-did-that-happen/

Whittaker, Maxine. "How Infectious Diseases Have Shaped Our Culture, Habits and Language." *The Conversation*. 2017. theconversation.com/how-infectious-diseases-have-shaped-our-culture-habits-and-language-75061

Wilson, Courtney. "Horses Succumbing to Ross River Virus in Queensland, Affecting Equestrian Events and Recreational Riders." *ABC News. Australian Broadcasting Commission*. 2015. Accessed 4 July 2019. www.abc.net.au/news/2015-05-16/increasing-number-of-horses-succumbing-to-ross-river-virus/6472620

Wright, A. E. *Mosquito Life Cycles, Ecology and Behaviour*. Ballina: Ballina Shire Council, 2002.

Wright, Sarah. "More-Than-Human, Emergent Belongings." *Progress in Human Geography* 39, no. 4 (2014): 391–411. https://dx.doi.org/10.1177/0309132514537132

Youle, Merry. *Thinking Like a Phage: The Genius of the Viruses That Infect Bacteria and Archaea*. San Diego: Wholon, 2017, Kindle.

5 Transgression and attunement

Clearly, all is not harmony and health within the relations between wallaby, mosquito, virus, and myself. Focusing upon these fractious relations helps to unsettle any too-eager movement between concepts such as co-becoming, kinship, and attunement, and discourses of harmony, balance, and flourishing. Uncritically linking relationality with these desired outcomes risks the assumption that merely *identifying* relationality might be a solution to the complex issues of anthropocentric destruction.[1] As Uli Beisel, Anne Kelly, and Noémi Tousinant note: "Focusing on the creatures that trouble coexistence opens up different registers of human-nonhuman interaction to scrutiny".[2] This chapter focuses on the problematic, the transgressional, the disruptive, and the ethically challenging within relational narratives. It follows the processes that occur as the mosquito pierces my skin, inserts its mouthparts, and takes my blood. Issues of boundary integrity, fear, and the role of threat in adaptive attunement are considered. Idealisations of harmony, balance, and well-being are examined, as I argue that ethical action cannot escape, but can engage with, the uncertainties inherent within the hermeneutic problematics of material semiotics.

Transgression

Transgression, pain, and pre-emption

It's a warm summer evening and I'm at my desk, a couple of kilometres from the Georges River. My ears are receiving many sounds; the rushing sound of wheels on asphalt, the refrigerator occasionally whining, lorikeets chattering. None of these sounds disturbs me. I am working, and they barely touch the edges of my awareness. Their vibrations move my eardrum, but somewhere decisions are being made to repress my alarm responses. Then, as a delicate, high pitched buzz sounds somewhere in the room, I stiffen and grow alert. The sound comes close by my ear, and I am distracted from my thoughts. I find my eyes darting around the space of the room, searching for a fragile, black, flying creature. The sound becomes quieter, and I relax a little. When

DOI: 10.4324/9781003286493-8

she lands on my ankle, her long, lithe legs spread her weight, so the sensitive nerves near my skin's surface do not detect her. Her mosquito proboscis is a highly specialised mouth comprising six separate needle-like stylets.[3] Two of these have cutting teeth to pierce through the skin, and another two are specialised to hold the skin tissue open. Saliva, containing substances that thin my blood, is injected through the fifth and the sixth is the tube which will carry my blood back to the mosquito's digestive system. The mosquito's mouth delicately slices into the skin and searches nimbly through the skin cells, following the taste of blood, searching for a blood vessel to puncture and enter. When my ankle itches, I feel she has outwitted me. As I scratch, the vague fear of Ross River Virus crosses my mind, in the form of the memory of a map I was shown of its prevalence around this region.

In a gesture of both redress and complication, I want to consider narrative strategies that not only contribute towards attunement and intensity but also necessarily involve harm and defer narratives of increasing individual security or well-being. *Aedes vigilax* invades across the boundary of my skin, a piercing that is commonly presented as an attack.[4] In transgressing my skin and removing parts of my body, she enacts another apparent transgression – theft. Val Plumwood considers a dominant trope in Western traditions is to regard the body as property that ought not to be shared unwillingly. She writes: "In the individual/justice universe you own the energy volume of your body absolutely and spend much of that energy defending it frantically against all comers." As an alternative, she proposes a "Heraclitean universe, [where] being in your body is more like having a volume out from the library, a volume subject to more or less instant recall by other borrowers – who rewrite the whole story when they get it".[5] Thinking about a body as property treats skin as a wall, the self as sovereign, and removes porosity and collaboration, for fear of the loss of what has been achieved in the becoming of the fortress being.[6]

While Plumwood's critique questions the human determination to maintain exclusive possession of the body as property and highlights the critical role that transgressions of boundaries play within their becomings, she underplays or ignores the importance of concomitant efforts to defend against violation within involutions of attunement and irritation. In acknowledging the importance of openness to being affected, we should not deny the performances and narratives of defence and exclusion, as Eva Haifa Giraud notes.[7] It is only one side of the coin to recognise that all of us who are aligned within co-oriented obligations benefit from the strengthened inter-connected responsibilities achieved from each of us remaining vulnerable and interdependent. The other side is a recognition that there are, and will continue to be, desires and processes working against these vulnerabilities, towards separation, defence, and security.[8] In the discussion ahead, I am interested in discourse between these trends, discourses such as that of pre-emption, that complicate this between.

Affect, including sensing, is a type of change. Paraphrasing Alfred North Whitehead, Steven Shaviro puts it like this: "The way that I receive a perception [...] is the way that my body changes, or has changed [...] the way that the feeling entity is affected, or changed, is the very content of what it feels".[9] This is the unifying move that a semiotic material position facilitates. Perceptual meaningfulness becomes as the body adjusts to accommodate, resolve, or otherwise respond to the changes of affect. Indeed, as we take a semiotic material view, perceptual meaningfulness *is* the body's adjustment to the affect. A new affect disrupts before a body can adjust to it, before it becomes resolved into becoming meaningful. This may feel like disturbance, discomfort, or pain: "The newly visioned blind do not see things. They feel a pain in their eyes" is a startling example described by Brian Massumi.[10] Transgression interferes, causes a pain of disturbance, which enfolds within the becoming meaningful of relations, not necessarily towards a resolution into a state of non-pain. As within that between mosquito, virus, and mammal, a relation may continue to incorporate pain, the pain may increase, change, or it may become unsustainable.

Where interactions involve regular stress or pain, pre-emptive actions and structures may become folded into the materiality of relations, complicating them. Much contemporary discussion amongst humans about our relationships with mosquitoes is concerned with ways of acting pre-emptively to avoid or minimise impacts upon human health. In my mosquito encounter, my pre-emptive actions include attempting to brush away or swat the mosquito. Efforts to poison or drain the puddles of water, employ nets or screens, or genetically engineer a population of mosquitoes are also pre-emptive. At a biologically adaptive level, we are pre-emptively becoming bodies that are startled by the sound of a mosquito. Pre-emption is folded into the attunements of senses. The relationship between fear and pre-emption includes a kind of self-perpetuating reflexive folding.[11]

Pre-emption and fear also fold out across geographies as "landscapes of fear". This refers to a landscape as experienced by prey species as variously coloured by degrees of safety and fear by the presence of a predator species.[12] Such a landscape is readable and nuanced in terms of fear: in this location, one must be alert and quiet, or only venture in twilight, another place, a burrow perhaps, permits ease and rest. The landscape of fear helps guide pre-emptive choices and behaviour. Inspired by Aldo Leopold's 1949 observations of the devastating changes to mountain ecologies when wolves were removed and deer, losing their fears, overgrazed the vegetation,[13] numerous rewilding projects have documented the dramatic changes in the behaviour of prey animals as their predators are re-established, and they read their landscapes as more emotionally nuanced. Within the Georges River catchment, the mosquito provides me and perhaps the wallaby, with experiences of wariness, particularly on warm summer evenings. Pre-emption and landscapes of fear also play out within a body's immune system, as I will discuss later in the chapter.

The kiss of life and the parasitic suck

Compared with the narrative of transgression and theft outlined above, Deborah Bird Rose provides a very different take on the piercing of boundaries as one creature enters inside another. She follows the blossoming eucalyptus tree, as it opens its nectar-rich flowers to the atmosphere and is met by the long and eager tongues of flying-foxes. Instead of a transgression: "Think of that meeting place where the tongue laps up the nectar, and where the tree opens its blossoms. Just there, at that meeting of tongue and blossom, life happens: there is an interspecies kiss of great eloquence",[14] which she describes as a "kiss of life".[15] The kiss she describes is chiasmic: inside it, edges are inter-responsive, but two stories remain distinct. Across communicative membranes, much-desired gifts are offered and received. The two stories entwine, shaping and obliging each other, then move off along their different unfoldings.

The similarities and differences between this interspecies boundary crossing and the one occurring across my boundary as the mosquito bites are worth comparing. The two encounters are alike in many ways. Two beings entangle bodily and behaviourally, so that for each, within the encounter, generations of accommodations to meet each other's fluidities are called into performance. In both encounters, nourishing fluids are exchanged across the boundary. Standard accounts, however, describe these encounters as belonging to two very different genres: between the flying-fox and the eucalyptus there is symbiotic mutualism, between the mosquito and me there is parasitism. Within the former relation, each organism performs important work to enable the life of the other. The symbiotic kiss enhances the lives of each being: the fox receives food; the tree is pollinated. Each species has manoeuvred the interaction towards its advantage. Within the parasitic relationship, only one partner benefits at the other's expense. The mosquito is regarded as strategically adapted to exploit me as a resource, while my body is not strategic in attracting the mosquito,[16] and I have not or at least not fully adapted myself to benefit from this visit because it potentially sickens and weakens me.

From this, I wish to draw out two discussions. First, to again challenge the too-rapid linking of relations and ethics and draw out a more nuanced account of relational ethical criteria based upon the folded semiotic material ontology developing through this book; second, to explore what happens if we complicate these two similar but different encounters. Rose muses that:

> From creature to creature, flying fox to tree, and from creature to atmosphere, tree to air, and from atmosphere to creature, oxygen to all who breathe: this is a cosmopolitics of togetherness produced through heterogeneous ways of being amongst creatures who need each other, who seduce and sustain each other. There is a foundational ethics within this multispecies kiss.[17]

While we are all complexly interdependent, among the innumerable encounters of cosmopolitical togetherness there are many which are more like vampiric sucks or viral sabotages than kisses of life. Any attempt to locate "a foundational ethics within a multispecies kiss" needs to take this into account, not least when the vampire is human or economic. The most common examples in the literature, the wasp and the orchid or the bee and the flower, portray the knots of interrelatedness as pleasant and contributing towards mutual flourishing.[18] But the relations between a mosquito, wallaby, virus, and human body are also reciprocally bound along many intertwisted avenues of evolutionary unfolding. Perhaps overall or in the long run, even here the outcome is better health and longevity for each of our four unfurling threads of life, but who could account with balance sheets the cost in terms of sickness, death, and suffering? Such a reckoning always involves the selection of one scale or layer to claim as the benchmark. From an ontological position that adopts fractally folded relations (Chapters 1–3 contrast layered and folded semiotic material ontologies), it makes no sense to ask about such ultimate outcomes, because every desired relational outcome already folds into others, as others fold into it. Criteria for the assessment of integrity or health are being made and revised within the process, along all the edges of responsive affect. Criteria are of the folding process. They are immanent/emergent in the turns which bring the reckonings of different perspectives (which can also be regarded as different scales, because of their different grounds for reckoning) to bear upon each other. I illustrated this in Chapter 4, through the example drawn from Rose's work, where criteria for the lively attuning of the flying-fox and the eucalyptus blossom shimmer at the interface between synchronous and sequential temporalities that bring their differently scaled processes of reckoning to bear upon each other. This means that from a folded ontological perspective, ethical discussion focuses upon the interpretative, evaluative, and political work involved in *how* relational stories and the practices that write and interpret them are produced, controlled, and reproduced.

The honeyed kiss of symbiotic mutualism and the vampiric suck of parasitism are relational stories that are strongly associated with "good" and "bad". However, the stories of these relations only hold from particular perspectives. From other perspectives, those same interactions may read as very different stories. Somewhere deep within every honeyed kiss, there are likely to be captured prisoners, parasites, and excluded others. For example, producing the copious nectar that attracts flying-foxes demands much of the resources of a tree, diverting energy from the tree's other needs, and those of its other users. On the other hand, the story of the virus, wallaby, mosquito, and me ceases to be a story about parasitism when considered more broadly. It expands to include all the predators whose flourishing depends upon an adequate supply of mosquitoes and mosquito eggs and larvae, and the creatures which in turn feed upon them. Together we may become a complex life-enhancing story, within which pain, death, and parasitism become parts of a broader narrative. Neither original reading can be claimed as correct nor can one at a larger scale override the more intimate reading. The perspective of the mangrove and the perspective of the parasite's host cannot be traded off against each other except by a violence that denies one perspective in favour of the other. Such acts of violence will continue to be perpetrated, but it is the challenge of ethical, relational narrative work to struggle to justify these trades and to avoid making them presumptively.

In my skin

Skin response

As saliva from the mosquito seeps around my skin cells, a wide range of immune cells begin responding.[19] Some immune system cells, particularly dermal mast cells, respond rapidly by moving to the feeding site and signalling for further immune processes.[20] Other immune cells begin releasing compounds that cause allergic and inflammatory responses, resulting in the well-known itchy red bump. Some of my other immune system functions, however, are impaired or disabled by the saliva.[21] The Ross River Virus (RRV) travelling within the saliva also quickly triggers my skin's immune system responses.[22] Keratinocytes are the most common skin cells of the human epidermis (the outer layer of the skin). Within these cells, viruses similar to RRV stimulate strong immune responses and fail to replicate and spread. But in others, such as the large Langerhans cells, which move around through the skin layers, these viruses can replicate. These cells then transfer the viruses through the skin layers to other parts of my body.[23]

Most of the time RRV exists as a single strand of genetic material, or RNA, 11,853 nucleotides long which includes codes for the production of four proteins. One of these proteins forms a viral coat, another forms a lipid layer around the coat, and the remaining two protrude from the surface.[24] When a viral envelope makes contact with one of my body's cell walls, the E1 glycoprotein reduces the PH of the cell surface, which creates an opening through which the virus can enter. Once inside my cell, if not disabled by immune factors, parts of the viral genome begin affecting the regular work of the cell, stopping or dampening it, while other parts get to work interacting with the cell's components to reproduce and then reassemble viral proteins.[25]

Viral landscapes of fear

Within the multitudes whose obligated orientations enact my body, the RRV virus provides a kind of priming for my immune system, disrupting it and keeping it alert, providing within my body something like the landscape of fear described previously. With experience, an immune system becomes a pre-emptive strategy figured within the collective obligations/desires of the body. If the members of a cell community orient towards a specific regularity of practice, a particular material semiotic narrative, a virus skews that orientation. Within the assemblage of the body, the virus stands in the position that Donna Haraway calls the coyote or joker, the disruptive factor that scrambles the emerging narrative and provokes the emergence of adaptability and diversity while risking dissassemblage. It nudges off kilter any forming stability and adds the random uncanny which is needed to prime the sensitivities of attunements, but in doing so, provokes discomfort. The microbiologist, Merry Youle, describes the disruptive action of phages as "always there – an ancient, dynamic, bubbling, creative force that underpins all life on Earth".[26] Haraway describes this kind of unstable narrative at the heart of semiotic materiality:

> The structure of the narrative generates its actants. In considering what kind of entity "nature" might be, I am looking for a coyote and historical grammar of the world, where deep structure can be quite a surprise, indeed, a veritable trickster.[27]

The narrativity underlying the relational, semiotic world includes elements that keep it askew.

The biopolitics of immunity, parasitism, landscapes of fear, and preemption are twisted and complicated.[28] We are implicated and becoming inside their processes where actions are always folding back into their own unfolding. Michel Serres is among those writers who celebrate the resilience that develops through increased immune system adaptability:

> The parasite gives the host the means to be safe from the parasite. The organism reinforces its resistance and increases its adaptability. It is moved a bit away from its equilibrium and it is then even more strongly at equilibrium. The generous hosts are therefore stronger than the bodies without visits.[29]

Human immunity is indeed highly adaptive, and bodies that have been exposed to parasites and had the opportunity to developed immunity may be more resilient than others in many circumstances. However, there is an additional element Serres does not consider. He connects interdependence and resilience at the scale of bodies, while ignoring the interactions occurring at other scales. Considering illness principally as the worthwhile costs of resilience at the individual level evaluates short-term pain in relation to a somewhat longer term gain and assumes a goal-oriented model of life, where the goal is individual or species longevity. As Youle points out in her discussion of the very lively arms race (as she describes it) between bacteria and phages, prey is never likely to desire becoming food.[30] Those who suffer in fear will not appreciate the greater good their suffering might be measured as contributing towards, as storied from a different scale of abstraction.

Ontological models that incorporate a nested hierarchy of levels can support the evaluative assumption that stability at higher levels is worth more than that at lower levels. Such an assumption occurs when the provision of efficiencies of scale is valued at the expense of relational narrative complexity and attentiveness.[31] This work calls for a reconsideration of this assumption as a matter of concern. Efficiencies of scale are arguably important in the diversification of life forms and ecologies, but their achievement is worked through relations, relations which matter, which are themselves matters of concern. Levels and efficiencies are achievements, gained through struggle, play, experimentation, and attentiveness. The relations involved can be nourishing or damaging, strategic or playful – and my point is that how they are matters.

When Haraway describes Trin Minh Ha's notion of being inappropriate/d as "not to fit the taxon, to be dislocated from the available maps specifying kinds of actors and kinds of narratives, not to be originally fixed by difference",[32] she raises it as a narrative tool that troubles the narrative form itself. The mosquitoes birthed from the mangroves of the Georges River pose this kind of critical deconstructive relationality to the humans living in the region, as the virus does within bodies. The mosquito veers into human (and presumably it is similar for wallaby) lives, into whatever nicely or sketchily constructed narrative we each might be supposing for ourselves and disturbs it. For most of us that disturbance is minor – a surprisingly distracting alert to that tiny sound and a demanding itch – for others it is more disruptive. We are put on a lean; we veer off course. My immune system and my habits change, and there are economic and social impacts. I move indoors at dusk, purchase and apply repellents, wear long sleeves and long trouser legs, and install screens on my windows. This mosquito does not fit the taxon in that she is not like other insects for me. She intrudes inside my body and into numerous laboratories where humans encroach deeply within her biochemistry and her genome. She brings the neglected and contested margins of insect/human encounters into central focus. Haraway and Mihn Ha propose that this is the role of an inappropriate/d other. These veers and leans irk the story and keep it intelligent or alert.

Sickness

After moving through my skin layers, RRV penetrates my skeletal muscle cells and replicates there before entering my blood from where it can move to other tissues throughout my body.[33] For seven to nine days,[34] the virus incubates within my cells, replicating and spreading before I become aware of any symptoms. After this period, I begin feeling unwell, developing rashes on my limbs and body and painful joints in my fingers, wrist, ankles, and knees, which grow red and swollen. I feel feverish and tired.[35]

I'm often struck by the poor health of the mangrove trees along parts of the Georges River. Walking along the Lime Kiln Bay mangrove boardwalk,

I seldom find a leaf within my reach that is not pocked with galls, desiccated by beetles or larvae, sunburnt or twisted. I see almost no mature tree in this mangrove forest whose trunk and branches are intact. All seem wracked with rot and injury, misshapen, with dead and fallen branches; they resonate with Haraway's "porous tissues and open edges of damaged but still ongoing living worlds".[36] Death is prominent here, being transformed everywhere into mangal life. Mangrove mud is full of small dead things, disintegrating shards of leaf and branch, small, curled beetle carapaces, fragments of shell and bone. All sink together and lose their separate energies into a cold mud mass. This mangal is visibly a place where life/non-life/death boundaries are muddied. Life is always moving towards non-life, and non-life moving towards life. A sick tree becomes, not a dead tree, but a becoming-other kind of life. There is no gap between sick tree and fungus. The living sick tree is already mostly fungus. The living sick fish is already mostly bacteria.[37] What passes for life here is not just alive, it is also a bit dead, and a bit non-living.[38] Viral replication also diffracts the life/non-life binary question, revealing it as inappropriate for some biological contexts. Like other binaries brought into question by relational and material semiotic approaches, the life/non-life binary becomes repositioned as contextual.

Reconceiving the self as an entangled multiple, sickness and health become a shifting negotiation of orientations. Behind the idea of sickness are assumptions about closures, durations, and relational stories.[39] What comprises my health changes with perspective, raising perplexing questions. The next section considers the contributions towards my body's functioning of viral RNA within my cells, raising the question of the extent to which the RRV and its cellular affects should be considered part of the broader experiment of a future-becoming that enfolds both our stories.

Attunement

Attunement

In the context of the semiotic materialism outlined in this book, attunement can be considered as the capacity for meaning to be shared within a relation. The flying-fox and the blossom, the wasp and the orchid, the rabbit and the fox (the second and third examples were featured in previous chapters) are said to be attuned to the extent that their semiotic material communications align in interaction. Within relations, attunement is always present in some sense because to be discerned, the narrativity of a pattern of relationality will have resonance for the sensual and perceptual and narrativising capacity of some perspectival nexus (this is another way to think of the insight of Deleuzian virtuality, which was discussed in Chapter 3). As the diverse schemas emerging through different interacting perceptual frameworks enfold and affect each other, rich and unstable material semiotics develop, change, overlap, at times collapse.[40]

Along the Georges River, conversations between *Aedes vigilax*, RRV, wallaby, and human are occurring, within the semiosis of our bodies. Our various bodies are attuning to each other. My hearing is primed to hear the mosquito's whine, faint though it is. She is attuned to the chemicals from my body, the carbon dioxide I exhale, and other chemicals I exude. My skin, with its immunological capacity, is prepared to receive her entry. Her mouth is ready to pierce and search, her saliva ready to numb, her eggs are ready to be nourished with my blood. The attunement within this relationship is sensitive and nuanced. We are variously attuned to skin, sight, hearing, smell, mouthparts, digestion, immune system, and flight habits. The RRV virus has also attuned over time to occupy its complicated niche. To move between bodies using a third, moving through and drawing from each one, the virus is like a fold passing through the relations: a viral wave washing through our three animal bodies. To have become such a thing that can do this requires time and experimentation.

Research indicates that somewhere between 40 and 80 per cent of the genetic material within the human genome may be of ancient viral origin.[41] This DNA (the gene sequences written into our chromosomes) is thought to have entered with viruses which, instead of killing the host cells, integrated viral RNA nucleotide sequences (gene sequences that can travel independently and then transcribe into DNA, and replicate from it) into host DNA. A considerable amount of human genetic makeup may thus have (generally at some time in the distant past) jumped across from another living thing as a virus, possibly with the aid of a vector such as a mosquito, rather than evolving slowly in situ. Current research is discovering many processes within our bodily functioning that are influenced by this viral-origin genetic material. It plays a role in placental functioning, during early foetal development, and in our immune system. It is also important in brain functioning where it may explain variations in adult cognitive function,[42] including some aspects of autism and schizophrenia,[43] and essential aspects of brain evolution.[44] Elissa Pastuzyn et al. summarise that viral mechanisms "have been intricately involved in the evolution of synaptic plasticity and animal cognition".[45] Mammalian cognition may be among the many animal characteristics whose development has been influenced by viruses.

We are attuned, but this is not a simple harmony, and it is not steady or stable. Nor, however, is it merely a relationship of enemies alert to each other because we are in a clash of eternal battle, as the human and mosquito relationship is so frequently characterised throughout popular and even scientific literature.[46] The work of attunement is ill-served by being described by such singular, simple narratives. Mosquito body and behaviours, viral processes, wallaby body, and my body are engaged in ongoing affect and response together. We are continually attuning semiotic material patterns. But none of us is oriented towards the others to further their welfare, nor to further the welfare of us all as an entangled being. There is no necessary orientation towards a common well-being. Most of us are, ostensibly, harmful to each other and yet, still, we are attuned. Attunement does not come from a desire for harmony or stability or sustainability. The continuing patterns may include continuing pain and sickness. As Haraway writes: "It's not a 'happy ending' we need, but a non-ending".[47]

How then, in the constant wash of non-endings and chronic multiplicity, can any narrative be claimed to have explanatory value? How then can there be any meaningful relationality? It is the same question as that encountered for systematicity and emergence. The problem does not need revision or answering so much as continual restating with the "how" held open. "How" is the call for approaches that "stay with the trouble" or inhabit the thick gap. The ongoing negotiations of complex body structures like sense organs, mouthparts, feeding behaviours, and immune systems among the wallaby, the mosquito, the virus, and me can be storied as relations in multiple ways. Concern for *how* is the invitation of a material semiotic approach. The next chapter will engage further with this invitation.

Notes

1 This point has been recently argued by Eva Haifa Giraud, *What Comes after Entanglement?: Activism, Antrhopocentrism, and an Ethics of Exclusion* (Durham and London: Duke University Press, 2019), Kindle.

2 Uli Beisel, Ann H. Kelly, and Noémi Tousignant, "Knowing Insects: Hosts, Vectors and Companions of Science," *Science as Culture* 22, no. 1 (2013): 9.

3 Gabriela Quirós, "How Mosquitoes Use Six Needles to Suck Your Blood", Deep Look, *KQED Science, KQED Inc.*, 2016, accessed 4 July 2019, www.kqed.org/science/728086/how-mosquitoes-use-six-needles-to-suck-your-blood

4 Margrit Shildrick writes, "the skin is both the limit of the embodied self and the site of potentially transgressive psychic investments. In consequence, any compromise of the organic unity and self-completion of the skin may signal monstrosity", in *Embodying the Monster: Encounters with the Vulnerable Self* (London and Thousand Oaks: Sage, 2002), 51, Kindle.

5 Val Plumwood, *The Eye of the Crocodile* (Canberra: Australian National University E Press), 2012, 33.

6 Part 3 of this book discusses the attractiveness and implications of walls and sovereignty.

7 Giraud, *What Comes after Entanglement?*.

8 Open invitations to trauma are unfortunately also politically prone to being exploited, a point also made by Giraud.

9 Steven Shaviro, *Without Criteria: Kant, Whitehead. Deleuze, and Aesthetics. Technologies of Lived Abstraction* (Cambridge: Massachusetts Institute of Technology Press, 2009), chapter 3, paragraphs 15–16, Kindle.

10 Brian Massumi, *Parables of the Virtual: Movement, Affect, Sensation* (Durham and London: Duke University Press, 2002), 160.

11 Brian Massumi argues that within the logic of pre-emptive action, there is an incipient requirement that fear becomes a lived constant, a state of anxious anticipation within our bodies, in "Fear (the Spectrum Said)," *Positions: East Asia Cultures Critique* (2005), https://muse.jhu.edu/article/185296

12 John W. Laundré, Lucinda Hernández, and William J. Ripple, "The Landscape of Fear: Ecological Implications of Being Afraid," *The Open Ecology Journal* 3 (2010): 1–7.
13 Aldo Leopold, *A Sand Country Almanac* (London and New York: Oxford University Press, 1949).
14 Deborah Bird Rose, "Cosmopolitics: The Kiss of Life," *New Formations* (2012), https://dx.doi.org/10.3898/NEWF.76.07.2012
15 Rose, "Cosmopolitics: The Kiss of Life".
16 This is an open question, as "I" benefit from the highly adaptive human immune system, not only as an individual body but as a member of the species which has been so enabled by it.
17 Rose, "Cosmopolitics: The Kiss of Life", 110.
18 Timothy Morton makes the point that entanglement between the bee and the flower is not the result of goal-oriented strategy, but of capture, as discussed in Chapters 3 and 4 of this book.

> The useless beauty of a flower is thus not a cynical ruse to make more plants. It's a viral cipher that serves no purpose, but that, when caught in another system, say a bee's search for nectar, ends up ironically reproducing itself,

he writes, in *Dark Ecology: For a Logic of Future Coexistence* (New York: Columbia University Press, 2016), chapter 3, paragraph 93, Kindle.
19 Laurence Briant, Philippe Desprès, Valérie Choumet, and Dorothée Missé, "Role of Skin Immune Cells on the Host Susceptibility to Mosquito-Borne Viruses," *Virology* 464–65 (2014): 26–32.
20 Briant et al., "Role", 29.
21 Ibid., 30.
22 Jade Redfern, "Differential Profiles of Mammalian and Mosquito Derived Ross River Virus" (University of Canberra, 2015), 25.
23 Briant et al., "Role", 27.
24 David Harley, Adrian Sleigh, and Scott Ritchie, "Ross River Virus Transmission, Infection, and Disease: A Cross-Disciplinary Review," *Clinical Microbiology Reviews* 14, no. 4 (2001): 911, https://dx.doi.org/10.1128/CMR.14.4.909-932.2001
25 Redfern, "Differential Profiles", 10–12.
26 Merry Youle, *Thinking Like a Phage: The Genius of the Viruses That Infect Bacteria and Archaea* (San Diego: Wholon, 2017), Preface, paragraph 6, Kindle.
27 Donna Haraway, "The Promises of Monsters: A Regenerative Politics for Inappropriate/D Others," in *Cultural Studies*, eds. Lawrence Grossberg, Cary Nelson, and Paula A. Treichler (New York: Routledge, 1992), 11.
28 For an overview of biopolitical accounts of immunity, see Nik Brown, *Immunitary Life: A Biopolitics of Immunity* (London: Palgrave Macmillan UK, 2019), https://doi.org/10.1057/978-1-137-55247-1
29 Michel Serres, *The Parasite*, Minneapolis: University of Minnesota Press, 2013, The Best Definition, paragraph 14, Kindle.
30 Youle, *Thinking*.
31 I have discussed the grip of this assumption on material semiotic ontology in previous chapters.
32 Haraway, "Promises", 69.

33 Harley, Sleigh, and Ritchie, "Ross River", 912.

34 7–9 days is the most commonly reported incubation period, but there are records at 3 days and 21 days, at the extremes, ibid., 913.

35 Ibid.

36 Donna Haraway, *Staying with the Trouble: Making Kin in the Chthulucene* (Durham and London: Duke University Press, 2016), 41.

37 Andreas Weber describes a sick hawthorne tree that "is beginning to become even more itself by becoming lively, a hub of activity, a gift that enriches others", in *Matter and Desire: An Erotic Ecology*, trans. Rory Bradley (White River Junction: Chelsea Green Publishing, 2017), 197, Kindle.

38 "Compost politics, then, is a complex dance balancing the forces of life, growth, and care on the one hand, and death, decay, and dissolution on the other hand—at the same time and in the same space", write Franklin Ginn, Uli Beisel, and Maan Barua, "Flourishing with Awkward Creatures: Togetherness, Vulnerability, Killing," *Environmental Humanities* 4 (2014): 121.

39 Ginn, Beisel, and Barua, "Flourishing with Awkward Creatures" 120; Vicki Kirby, "Transgression: Normativity's Self-Inversion," *Differences: A Journal of Femininst Cultural Studies* 26, no. 1 (2015): 96–116.

40 For a discussion of affect and the enfoldedness of perception see Part 1 "Strange reflexivities".

41 Rafi Letzter, "Ancient Virus May Be Responsible for Human Consciousness," *Livescience*, 2018, www.livescience.com/61627-ancient-virus-brain.html; Nicholas Parrish and Keizo Tomonaga, "Endogenized Viral Sequences in Mammals," *Current Opinion in Microbiology* 31 (2016), https://dx.doi.org/10.1016/j.mib.2016.03.002

42 Luis de la Torre-Ubieta et al., "The Dynamic Landscape of Open Chromatin During Human Cortical Neurogenesis," *Cell* 172, no. 1 (2018), https://dx.doi.org/10.1016/j.cell.2017.12.014

43 James Ashley et al., "Retrovirus-Like Gag Protein ARC1 Binds RNA and Traffics across Synaptic Boutons," *Cell* 172, no. 1–2 (Jan 2018), https://dx.doi.org/10.1016/j.cell.2017.12.022

44 James A. Briggs et al., "Mechanisms of Long Non-Coding Rnas in Mammalian Nervous System Development, Plasticity, Disease, and Evolution," *Neuron* 88, no. 5 (Dec 2015), https://dx.doi.org/10.1016/j.neuron.2015.09.045

45 Elissa D. Pastuzyn et al., "The Neuronal Gene Arc Encodes a Repurposed Retrotransposon Gag Protein That Mediates Intercellular RNA Transfer," *Cell* 172, no. 1–2 (Jan 2018): 277, https://dx.doi.org/10.1016/j.cell.2017.12.024

46 See, for example, Youle, *Thinking*. This trope is discussed further in the next chapter.

47 Haraway, "Promises", 327.

References

Ashley, James, Benjamin Cordy, Diandra Lucia, Lee G. Fradkin, et al. "Retrovirus-Like Gag Protein ARC1 Binds RNA and Traffics across Synaptic Boutons." *Cell* 172, no. 1–2 (Jan 2018): 262–74.e11. https://dx.doi.org/10.1016/j.cell.2017.12.022

Beisel, Uli, Ann H. Kelly, and Noémi Tousignant. "Knowing Insects: Hosts, Vectors and Companions of Science." *Science as Culture* 22, no. 1 (2013): 1–15. https://dx.doi.org/10.1080/09505431.2013.776367

Briant, Laurence, Philippe Desprès, Valérie Choumet, and Dorothée Missé. "Role of Skin Immune Cells on the Host Susceptibility to Mosquito-Borne Viruses." *Virology* 464, no. 65 (2014): 26–32. https://doi.org/10.1016/j.virol.2014.06.023

Briggs, James A., Ernst J. Wolvetang, John S. Mattick, John L. Rinn, et al. "Mechanisms of Long Non-Coding RNAs in Mammalian Nervous System Development, Plasticity, Disease, and Evolution." *Neuron* 88, no. 5 (Dec 2015): 861–77. https://dx.doi.org/10.1016/j.neuron.2015.09.045

Brown, Nik. *Immunitary Life: A Biopolitics of Immunity*. London: Palgrave Macmillan UK, 2019. https://doi.org/10.1057/978-1-137-55247-1

de la Torre-Ubieta, Luis, Jason L. Stein, Hyejung Won, Carli K. Opland, et al. "The Dynamic Landscape of Open Chromatin During Human Cortical Neurogenesis." *Cell* 172, no. 1 (2018): 289–304. https://dx.doi.org/10.1016/j.cell.2017.12.014

Ginn, Franklin, Uli Beisel, and Maan Barua. "Flourishing with Awkward Creatures: Togetherness, Vulnerability, Killing." *Environmental Humanities* 4 (2014): 113–23.

Giraud, Eva Haifa. *What Comes after Entanglement?: Activism, Antrhopocentrism, and an Ethics of Exclusion*. Durham and London: Duke University Press, 2019. Kindle.

Haraway, Donna. "The Promises of Monsters: A Regenerative Politics for Inappropriate/D Others." In *Cultural Studies*, edited by Lawrence Grossberg, Cary Nelson, and Paula A. Treichler, 295–337. New York: Routledge, 1992.

———. *Staying with the Trouble: Making Kin in the Chthulucene*. Durham and London: Duke University Press, 2016.

Harley, David, Adrian Sleigh, and Scott Ritchie. "Ross River Virus Transmission, Infection, and Disease: A Cross-Disciplinary Review." *Clinical Microbiology Reviews* 14, no. 4 (2001): 909–32. https://dx.doi.org/10.1128/CMR.14.4.909-932.2001

Kirby, Vicki. "Transgression: Normativity's Self-Inversion." *Differences: A Journal of Feminist Cultural Studies* 26, no. 1 (2015): 96–116.

Laundré, John W, Lucinda Hernández, and William J Ripple. "The Landscape of Fear: Ecological Implications of Being Afraid." *The Open Ecology Journal* 3 (2010): 1–7.

Leopold, Aldo. *A Sand Country Almanac*. London and New York: Oxford University Press, 1949.

Letzter, Rafi. "Ancient Virus May Be Responsible for Human Consciousness." *Livescience*. 2018. www.livescience.com/61627-ancient-virus-brain.html

Massumi, Brian. "Fear (the Spectrum Said)." *Positions: East Asia Cultures Critique* 13, no. 1 (2005): 31–48. https://muse.jhu.edu/article/185296

Massumi, Brian. *Parables of the Virtual: Movement, Affect, Sensation. Durham and London*: Duke University Press, 2002.

Morton, Timothy. *Dark Ecology: For a Logic of Future Coexistence*. New York: Columbia University Press, 2016, Kindle.

Parrish, Nicholas and Keizo Tomonaga. "Endogenized Viral Sequences in Mammals." *Current Opinion in Microbiology* 31 (2016): 176–83. https://dx.doi.org/10.1016/j.mib.2016.03.002

Pastuzyn, Elissa D., Cameron E. Day, Rachel B. Kearns, Madeline Kyrke-Smith, et al. "The Neuronal Gene Arc Encodes a Repurposed Retrotransposon Gag Protein That Mediates Intercellular RNA Transfer." *Cell* 172, no. 1–2 (Jan 11 2018): 275–88 e18. https://dx.doi.org/10.1016/j.cell.2017.12.024

Plumwood, Val. *The Eye of the Crocodile*. Canberra: Australian National University E Press, 2012.

Quirós, Gabriela. "How Mosquitoes Use Six Needles to Suck Your Blood." *Deep Look* KQED Science. KQED Inc., 2016. Accessed 4 July 2019. www.kqed.org/science/728086/how-mosquitoes-use-six-needles-to-suck-your-blood

Redfern, Jade. "Differential Profiles of Mammalian and Mosquito Derived Ross River Virus." University of Canberra, 2015.

Rose, Deborah Bird. "Cosmopolitics: The Kiss of Life." *New Formations* 76 (2012): 101–113. https://dx.doi.org/10.3898/NEWF.76.07.2012

Serres, Michel. *The Parasite*. Minneapolis: University of Minnesota Press, 2013, Kindle. Originally published as *Le Parasite*, Grasset et Fasquelle, 1980.

Shaviro, Steven. *Without Criteria: Kant, Whitehead. Deleuze, and Aesthetics. Technologies of Lived Abstraction*. Cambridge: Massachusetts Institute of Technology Press, 2009, Kindle.

Shildrick, Margrit. *Embodying the Monster: Encounters with the Vulnerable Self. Published in Association with Theory, Culture & Society*. London and Thousand Oaks: Sage, 2002, Kindle.

Weber, Andreas. *Matter and Desire: An Erotic Ecology*. Translated by Rory Bradley. White River Junction: Chelsea Green Publishing, 2017. Kindle. Originally published as *Lebendigkeit: Eine Erotische Ökologie*, Kösel Verlag, München, 2014.

Youle, Merry. *Thinking Like a Phage: The Genius of the Viruses That Infect Bacteria and Archaea*. San Diego: Wholon, 2017, Kindle.

6 It matters what stories we tell to tell other stories[1]

The act of a mosquito bite brings together mosquito, virus, and human bodies: three very differently inhabited perspectives or Umwelts aligning over many generations. We are related through the processes of biting, infecting, feeding, and immune responses that confuse the distinctions between us. Certain ways and scales of reading of our connections have become highly influential, for example, understanding bodies as owned and defended, narratives of pre-emption, narratives of shared vulnerability within ecological systems, and the role of the virus and mosquito as the joker or troubler which, at some more extensive scale, may promote strength or resilience. In each case, the narrative perspective privileges one interpretive scale, and discounts others. Conceptualising relations as narratives of entangled obligations calls attention to how relations are storied and raises assumed scales of narrative privilege as a matter of concern. A material semiotic understanding of relationality draws attention to how relations are written, read, interpreted, recounted, transformed, remembered, interwoven, and shared, taking the texts as lived relations, as well as accounts of them. This is the area of interest described by Serenella Iovino and Serpil Oppermann as material ecocriticism, the critical analysis of material narrativity, but it is also about how we live relations.[2]

Misplaced hope in the relational turn

Hope in the relational turn

Many thinkers have turned towards relational approaches as a means to rethink responsibility.[3] The hope for many has been that attending to relations will help us to be more attentive, and hence more responsible, because rather than focusing on shoring up boundaries, we will instead focus upon the effects of our actions. Such hope can be read in statements such as "[e]thics is [...] about responsibility and accountability for the lively relationalities of becoming of which we are a part";[4] and "[k]in means cultivating response-ability for each other, whether one wants to or not. I have a relative; a relative has me. Let us recognize relatives as fast and well as 'we'

DOI: 10.4324/9781003286493-9

can",[5] and "participation in a world of cobecoming necessarily implicates us: insofar as we all help to shape worlds, we are accountable for how and what they become".[6]

Against the trends outlined above that seek to locate ethics within relational ontologies, Eva Haifa Giraud argues that such writings demonstrate "a tendency to celebrate entanglement – or treat it as a good in itself", assuming that more ethically sound actions "automatically proceed from the recognition of relationality", with insufficient attention given to exploring the details of how to intervene ethically.[7] Giraud's counter-move against these trends is to argue for the foregrounding of the constitutive and political roles of exclusion in any relation.[8] I suggest greater attention is paid to the material narrative strategies of relationality more broadly, including but not limited to, those involved with exclusion. I want to consider exclusion as one of the narrativising processes made among interactions that constrain and nurture relations into meaningfulness.

How well a turn towards a relational ontology can assist with informing ethics depends upon how relations are storied – in the material sense of this book. When Donna Haraway discusses the ethicality of kinship, she embraces an understanding of relations as storied in ways that seek to allow multiple articulations, that remain inadequate, and that incorporate reflexivity. These are the sorts of stories suggested by ecologies of practice and situated knowledges. But that is far from the case with all or even most accounts of relations. Haraway claims, "it matters what stories we tell to tell other stories with [...]. It matters what stories make worlds, what worlds make stories".[9] How relations are narrativised makes all the difference. If in turning to relations to avoid the closures that things impose, we impose the same closures upon relations or overlook the discursive practices within their storying, the move to relations may provide little guidance. Being storied means relations have writers, readers, reviewers, and translators who have situated and fluid perspectives. There are countless stories, innumerable ways to write them, and still more ways they interact. Relations exist in the processes of being written, interpreted, revised, repeated, and diffracted, whether by mosquitoes, viruses, wallabies, humans, or among us as we knot together. If relationality is to guide ethics, critical engagement with hermeneutic demands is needed.

When relations are discussed, they are often characterised within narrative categories. Relations with mosquitoes are frequently portrayed as being about human health or community hygiene, for example. The relation can quickly become fitted into a genre so it can be thought and managed. Characterising a relation within a narrative category – the mosquito's interaction with me as typically parasitic for example – can imply the relational story is known and manageable. It has been fitted to a mould and can then move, as one discursive object, relatively efficiently into further human-centred discursive contexts; a community meeting, a laboratory, a classroom, a backyard barbecue. Haraway demonstrates this point when she outlines the mechanisms through which the particular narratives of the "modest witness" of science have come to have the "magical power" of appearing to be "clear mirrors".[10] They have lost their history as stories within particular narrative lines, among others, and positioned themselves as the only story worth attention. Closing a relation story and maintaining that closure captures the particular discernments of those whose privileged positions have enabled them to reinforce one another to the point of becoming narrative institutions. Relationship is made of story, so it is also made of power, made of characterisations tied to frames of judgement. Nothing comes without its world, as Haraway tells us,[11] and it is the same with relationality, it comes with its world too, already entangled. The ethical limitations of thing-think are not escaped simply by swapping to relation-think; both require the same critical attention to privilege and power. Reducing relations to genre-based characterisations is a hegemonic practice, semiotic material theory supports critique and analysis of these practices.

War stories and the hegemony of genre

Mosquito-human relations are very often characterised as competitive and strategic, or weaponised and militaristic[12] and that between a virus and a host immune system is frequently written in either military language as an arms race between skilled and strategic adversaries, or as a "militarised automated factory".[13] In a reflection upon his own autoimmune-related health issues, Ed Cohen explores and critiques the conflict-based conception of self/non-self proposed by the standard version of immunity. The familiar version of the immunological self, according to Cohen, is "that which does not react to itself".[14] The standard narrative of the immune response then "relies on a logical and biological bifurcation of the world into self and not self" and posits "a fundamental hostility [...] between what [the organism] recognizes as itself and everything else... . autoimmunity constitutes a failure of defence "intelligence" and hence manifests an instance of 'friendly fire' ".[15]

As Cohen points out, the language of military metaphor has profoundly influenced the notion of self as read through the narrative of immunity and autoimmunity. These metaphors were adopted during the late nineteenth century as a response to political and economic problems arising because of quarantine restrictions and were later extended into the bioscientific literature. Conceptualisations of biological immunity inherited a colonial biopolitics concerned with justifying the exclusion of a privileged few from the quarantine obligations required of the commons.[16] These metaphors perpetuate assumptions about what is "natural" that are based upon one particularly political philosophical paradigm, that which privileges the divide between friend/enemy as definitive of the basis of self/non-self.[17] Cohen argues that there is no sound basis for the metaphoric transfer of this oppositional relationship from a human historical and political context to characterise interactions occurring within animal bodies. A growing knowledge of the extent of symbiotic and commensal relationships occurring within animal cells, organs, and bodies casts doubt upon any simple oppositional characterisation.

The relationships of beings living together within a body are complex. A story based upon one metaphor will not do to characterise them. Cohen suggests an alternative; that "perhaps what we now call biological immunity is at best an exceptional mode of biological community or conviviality".[18] Haraway offers a similar alternative story, a "network theory" model of immunology which locates all the interactions of an immune response within the immune system rather than dividing the responses between defensive host and offensive invader.[19] The immune system is thus "indefinite if not infinite", an "adaptable communication system with many interfaces" dispersed across many tissues and organs.[20] Its many components are continually making themselves more adaptable through constantly changing themselves to mirror each other. This behaviour results in increasing adaptability.[21] The immune system relations are storied as occurring within a continually self-challenging, self-educating system. Haraway extends the community of this system to include the scientists, infected humans, drugs, and medical technologies. After these revisions by Haraway and Cohen, the delicate, experimental equilibration processes of immunity are more about politeness than war strategy. With the immune system, Haraway finds: "A radical conception of connection emerges unexpectedly at the core of the defended self".[22] "Replaced by subtle plays of partially mirrored readings and responses, self and other lose their rationalistic oppositional quality", she concludes.[23]

The battle narrative not only misrepresents the complexity of interactions within immunity, but it also actively constructs boundaries and opposing sides to suit the maintenance of a particular status quo: the body at an apparent moment of purity. This narrative appears to be designed to justify holding in stasis that moment which is assessed as positioning the human storyteller to the most advantage while denying a narrative of interrelational communication and adaptive change. Nik Brown, interpreting the work of Mary Douglas and Julia Kristeva on purity and abjection into an immunity context, argues that "[p]urity's terminal motionlessness" in this self/other story of immunity is dangerous because it disables the mechanisms for adaptive change that occur with boundary crossings, and that would ultimately provide greater security into the future.[24] The attractions and consequences of narratives that still time to hold onto an apparent advantage will be discussed further in Part 3 in relation to the building of embankments and dams on the Cooks River.

Articulated stories

Relational stories expand or change with a changing perspective or with interpretation into a different context. Haraway uses the concept of articulation to describe the processes of communication among diverse members of networks working from different perspectives but oriented towards overlapping goals. "Discourse is only one process of articulation. An articulated world has an undecidable number of modes and sites where connections can be made", she explains.[25] To meet the needs of these broader networks, the expansive kinship communities and the other deeply entangled trajectories that Haraway describes; relations need to be storied in multiple, diverse, responsive ways. Ann Kelly and Javier Lezaun, researching community practices of mosquito monitoring in Dar es Salaam, provide a useful mosquito example.[26] They find it requires several different ways of storying to understand the idea of a mosquito breeding ground in the context of the city. To adequately understand the breeding ground for the purposes of living flourishing human lives among malaria mosquitoes in this city, more than one material semiotic narrative framework is needed.[27] Multiple and diverse lexicons and grammars are required in order to speak of the interactions between interdependent and inter-affecting beings. This example illustrates how unlikely it is that any one narrative or metaphor will be appropriate for the task. Seldom will it be most apt to identify only competition as the mode of interactive intent any more than only love or only care or only the desire to extend one's own existence.

Bruno Latour argues along these lines when he calls for de-identifying groups and categorised relationships and instead "follow[ing] the actors's own ways" into interactive spaces.[28] As research changes attention towards connections and associations, he suggests, the things themselves will retreat, and the connections become more vivid. For Latour a relation is a narrative that provides a meaningful history or account of how things matter to each other: "What circulates, so to speak, 'inside' the conduits are the very acts of giving something a dimension".[29] Following these insights, holding off on identifying relations is just as important as holding off on identifying bounded entities. Nominating "care" or "competition" as explanatory labels for fixing a relation as a well-honed narrative account is as prescriptive of relationship as is designating a being by the boundary of its skin. If we seek to push beyond prescriptive meaning, Latour, Haraway, and their fellow material semiotic thinkers suggest we should hold off on identifying or judging relations as being definitively of a type and remain wary of relational narratives that fall into well-honed genres. Narratives of relationality do not need to be told as closed or settled. They are poorly accounted for by imposing a singular, genre-limited unchanging story. The relationship between myself, the mosquito, the virus, and the wallaby is not only disruptive, it is not only regular, not only harmonious, not only a tale of flux. Relational narratives in the mangroves are often likely to be slippery and sticky, looping, multiplying, and passing through diverse bodies, sometimes many times over.

Narrative tools that foster attunement

Perspectival stories

There are very many ways to story and many different tools with which stories are written, all of which are, in their ways, stories too. These are creative processes of meaning-making, made within intra-affecting, fluid circumstances. Such creative processes can be more or less risk-tolerant, more or less oriented towards security or efficiency or accommodation or generosity or competition or justice (or other possible orientations stories can take). The storying may recognise its own guessing, experimenting, iterative processes of finding the best fit, or it may indeed involve some resistance to fitting, which also may be recognised. Alternatively, the discernment may oversimplify and fail to include a recognition of these iterative processes and take itself as a privileged position. Because material semiotic storying is something that is actively performed, the responsibility that is so often related to a shift towards relationality is perhaps more helpfully conceived as a responsibility for how the stories that construct relations are written and interpreted. There are narrative tools for keeping meanings open, for staying with more ambiguous, historied meanings, for storying porous boundaries, for avoiding fetishising the settled plots that genre suggests. Many of these narrative approaches are explored in creative works, or documented in anthropological, ecological, and organisational research. I will begin by considering perspectival ways of knowing, using Eduardo Viveiros de Castro's analysis as a guide.

Viveiros de Castro identifies certain cultures as perspectival, a term he uses in a specific way.[30] He finds that while everywhere there are multiple human and non-human perspectives, some particular Indigenous groups that he has studied prioritise maintaining awareness of this plurality. While other cultural groups presume one privileged perspective as the writers of the stories that truly explain the world, perspectival cultures attempt to keep open the realisation that, although not all perspectives will be accessible to the human group, they are present, storying in their own way, equal in value, and worthy of as much respect as can be achieved. Thus, within the perspectival framework, stories (Viveiros de Castro uses the term "myth" rather than "story") must recognise they are situated within a diversity of other stories. They must acknowledge that the meanings of these other stories are only partially (if at all) accessible and that their own stories are interwoven complexly, but not always harmoniously, with these others.

Perspectivism, as Viveiros de Castro describes it, seeks to retain multiplicity and works hard to not settle on a privileged perspective. There is no overarching structure to perspectival mythmaking: structures only exist as they emerge within the interpretive and transformative work occurring as differing perspectives seek to understand or accommodate or otherwise relate to each other. As Peter Skafish puts it, structures "are only found … between two variants, sequences, or levels of a myth inasmuch as they recast each other".[31] Additionally, perspectivism shows a way of storying relations and a relational ontology that is more about "'unreducing' and undefining them, by bending every line of division into an infinitely complex curve. It is not a question of erasing the contours but of folding and thickening them, diffracting and rendering them iridescent".[32] Perspectival cultures, according to Viveiros de Castro, work hard at storying methodology with the intent of maintaining an openness, within their storying practices, to acknowledging and allowing for the multitudes of storyings that are being written/enacted/performed/told from and within other perspectives.

Since relations are written/read from perspectives which may claim their privileged reading as reality, the question of whose perspective a story privileges is pertinent. Cohen, in his discussion of immunity, for example, points out there is a brutal colonial history behind the widespread employment of military metaphor for immune processes. He demonstrates how this particular language and storying framework has had significant feedback effects for popular and current stories about the nature of healthy bodies, individual physical integrity and treatment approaches.[33] For both Haraway and Deborah Bird Rose, this is an issue of concern. They both explore the way that power is entwined within relational stories that come to matter. Their analyses are not limited to tracing what the stories are, more important to both of them is understanding the ways these stories get told. For Haraway, much of this exploration centres upon narratives of science and technology and their broad influences upon human and more-than-human lives.[34] Rose focuses her analysis of power upon the ways of storying relations with country and with non-humans that have recently come to matter within Australian legal, economic, and political contexts. She contrasts these with particular Indigenous Australian more-than-human ways of storying that have not come to matter within those contexts.[35] Both thinkers are interested in exploring how it could be otherwise and approach this by putting the ways of storying first – emphasising that it is the ways we story, not merely the stories themselves that influence the shape power takes within relations. Haraway looks for alternatives from within feminist and Native American approaches, and from her experiences as a dog trainer. Rose turns to the lessons of her Yarralin and Lingara teachers from Northern Australia. For both Haraway and Rose, a key point is that writing the relational world takes attentive work, and that makes relations promising sites for far-reaching political impacts. Both consider it essential that we attend to discursive practices and hermeneutic concerns as sites of potential leverage for change within the semiotic worlds of relationality.

Kin and companion species

Both Haraway and Rose consider at length the ways that storying relations as kinship or companion species can be helpful for implementing more care between humans and non-human animals. Haraway suggests kinship provides a way of thinking that aligns with the objectives of multispecies environmental justice, which "means nurturing and inventing enduring multispecies – human and nonhuman – kindreds".[36] For Haraway, kin is an activity. It is something to decide to work at by engaging in particular communicative in-betweens in specific ways. She requires us to ask "What must be cut and what must be tied if multispecies flourishing on earth, including human and other-than-human beings in kinship, are to have a chance?"[37] As ethical tools, kin and companion species do not offer any easy release from cosmopolitical trouble: "Which companion species will, and should, live and die, and how" is always an open question, one continuously of concern.[38] She even suggests that "the dilemmas posed by blood-sucking insects" can help us to think about how the work of figuring responses within the dances of becoming-with in ways that "escape calculations" of the sort imposed by ready-made ethical narratives.[39]

When she introduced the concept of companion species, Haraway was thinking of her faithful dog, Ms Cayenne Pepper.[40] Cayenne and Haraway formed a team that trained and competed together; they read each other's signals and worked together as partners in their sport. They loved and longed and played together, sharing bodily fluids as long-tongued dog and human frequently do in contemporary cultures where dogs are maintained as house pets. Further, as domesticated dog and human, Cayenne and Haraway continued a long tradition of reading each other's signals, fitting into each other's needs and behaviours, dwelling alongside "in obligatory, constitutive, historical, protean relationships".[41] Haraway postulates that her immune system must have been changed through its engagement with Cayenne's sloppy kisses. Presumably, Cayenne's immune system also responded to Haraway's saliva, so that the relationship is physiological as well as emotional, behavioural, and sensual. Additionally, the relationship is cultural, dance-like,[42] embedded within human, dog, and human-dog practices and traditions of meaning-making and living; hence Haraway's use of the term "natureculture" to describe the space of companion species relationships.[43]

Like Cayenne Pepper and Haraway, virus, mosquito, wallaby, and human are calling each other into becoming who we are in response to each other physiologically, culturally, "all the way down [...] into naturecultures".[44] Mosquitoes are implicated in human bodies, decision making, domestic rituals, architectural design, land use management, emotional lives, family relationships, sleep habits, health practices, and policies. Humans are also implicated in mosquito bodies and their decisions. Mosquitoes in some locations are changing their behaviour as they adapt to insecticidal mosquito nets, feeding earlier in the evening and outside instead of inside houses. Further, mosquito bodies are continuing to find mechanisms to enable them to survive pesticides through building their immunity or through breaking down the poisons into harmless compounds.[45] But while we are often eager participants in our relations with dogs, humans often "remain hesitant about the possibilities of becoming-with insects",[46] knowing that "insect control and capture might be the price we have to pay to sustain our human existence".[47]

There are several issues Haraway's concepts of kinship and companion species raise as they filter through mosquito, virus, and human bodies. Considering humans and mosquitoes as kin or companion species reminds us that not all companions are welcome, and that the relationship is likely be regarded very differently by each partner. If kinship is something we do because it makes us better carers of each other, as it is for Haraway, what happens to the ethics when kinship involves mutual harm, as it often does between myself and the mosquito? Should I perhaps choose not to be kin with a mosquito, because the relationship is not mutually life-enhancing, and instead describe the relationship in another way – as competitive or parasitic perhaps? This is the position Uli Beisel takes when she interrogates the appropriateness of applying Haraway's concept of companion species to a malaria-carrying mosquito.[48] However "kin" does not seem helpful as an ethical guide if it is merely a justification for choosing only those who benefit me and excluding those I prefer to avoid. On the other hand, if I am obliged to be troubled kin with the mosquito, kinship opens into something so ethically expansive that it is difficult to find guidance towards caring from it. As the concept gains the universality it requires to become useful in a more-than-human context, it loses the capacity to support ethical implications.

Like Haraway, Rose also writes of interspecies kinship with dogs, in her case, dingoes. Her discussions of the ethics entangled in kinship emphasise the importance of holding off on directing or evaluating the stories of life, and on acknowledging the honest "I don't know" as a caring, ethical response.[49] "I don't know" is what her Indigenous teacher, Old Tom, an old man who lived among many dogs, carefully and repeatedly offered as the conclusion to a story about saving a drowning man, when asked about the outcome. Rose finds a profound insight in his answer. Old Tom says "I don't know" because this allows the stories of the world to unfurl without his disruptive meddling. When one is inside multiplicities of stories, entangled as threads among threads, and neither able to nor desirous of becoming untangled enough to step outside, Rose suggests that "I don't know" is an expression of the significance of that complexity and one's own entanglement within it. It can be a polite response,[50] in the way it enables and invites others to participate:

> In refusing an overt decision-making process, he can be seen to be asserting a kind of love – a faithfulness to life in which call-and-response are yet another flip. [...] The flip connects even as it differentiates. It is foundational to world making in a life-affirming awareness of uncertainty and connectivity.[51]

A matter of manners

Ethics folds into storying in the doing. If my version of the story of the relationship between myself, a wallaby, an *Aedes vigilax* mosquito, and RRV is written (which, from a material semiotic perspective, concerns how it is performed through bodies, behaviours, economically, and so on[52]) so that it breaks and pauses to allow scope for stories from other perspectives to develop, if it takes reflexive turns to notice its own narrative pattern, if it opens itself to consider the multispecies, historical, cultural, linguistic influences that contribute towards its complexities, if it refrains from construing itself as superior, all of these approaches within the way I story the relation will affect the way it can be materially storied by others. Ethics becomes located within the manner and manners of our storying.[53]

In her animal studies research in laboratories and field studies, Vinciane Despret demonstrates some of the complexities of the manners of multispecies relations in terms of politeness. She reveals, through many examples, animals attuning to researchers' or trainers' expectations and attempting to provide the expected responses, showing "animals and people have succeeded in becoming attuned to what matters to the other, to act so that what matters to the other also matters to oneself".[54] She describes this sensitivity in terms of the idea of politeness. Politeness involves not imposing one's own stories upon others. It means holding off from making assumptions, holding off on defining or constraining within predetermined expectations. It includes allowing the Other space and time to offer, in their own way, their suggestions for the shape and direction of the interaction, while showing interest, while demonstrating a desire that they do so. One of Despret's many examples involves a mouse experiment in which a polite scientist offers "the possibility that the animal could show how he takes a position with respect to what is asked of him".[55] Being polite, as Despret proposes it, functions as an enabler of attunement and agencement[56] – the increase of agential potential gained by mutually supportive entities.

By beginning from the observation that animals are already interested (to varying degrees and in diverse ways) and already practising their own politenesses in relationship, Despret suggests the world is ready, waiting only for our (human) politeness in return.[57] She suggests that we embark into relationship without expecting or demanding a particular response, holding off to allow the unfolding relational pattern more potential to experiment its form. This may require some false starts, some backtracking, some embarrassment, some risk. Each relationship is more dangerous as I leave it open, compared with closing it off to maintain my simple short-term advantage. Killing the mosquito before it bites me, for example, is safer than waiting to see whether I develop RRV. Safer, that is to say, in relation to the predictability of the future for the body as it is defined within the time and space trajectory contained within my skin. If I kill, I will not become ill, but also will not develop immunity or lessen my allergic reaction to the next bite,[58] and the mosquito and virus will not have the opportunity to adapt to my body. If I abstain from killing and am bitten, I may develop immunity and reduce the intensity of my future allergic reactions, I may become ill, or both. The mosquito and virus will continue to adapt to my body. Potentially the virus may incorporate into my cells in ways that change me significantly, and my descendants even more. Whether or not I kill the mosquito impacts upon the playing out of many future attunements.[59]

I wonder what it means to compare Despret's notion of politeness with Haraway's call to allow space for being inappropriate/d? Humility and awareness, taking the time to hear, to be open to the other are the hallmarks of politeness. Haraway's inappropriate/dness, in contrast, comes at things askew and skews them in turn. This might seem a direct contrast to politeness, but in a world of relationships, twists are woven into the fabric of affect and prehension. Politeness is necessary because of the skew. It is because we are askew; ignorant and operating within environments that are opaque, within relations that are like "a multipartner mud dance",[60] that we need to pay acute attention to communication and interpretation. Inheritance and adaptation, culture and semiotics, are all emerging stories among the microorganisms of my body and those of *Aedes vigilax*, as between us and among us with other mangrove contributors. Haraway appreciates this extended social: "We are all in chiasmatic borderlands, liminal areas where new shapes, new kinds of action and responsibility, are gestating in the world".[61]

Arts of noticing and interpreting

Writing and reading are not two separate activities; they are the one thing, chiasmatically together. Strategies for telling stories in ways that disperse and delay, fragment and allow for new and experimental connections, must team with strategies for reading/listening/paying attention and interpreting that can attend to the demands of these knotted and multiple-threaded texts. "We know that nonhumans communicate in multiple registers, and perhaps it is necessary, therefore, to be able to listen in multiple registers", Rose writes.[62]

Anna Tsing turns to the autonomous intertwining melodies of polyphonic music to illustrate a certain kind of noticing that demands concerted, attentive work. Polyphonic music requires active and demanding listening, so in listening to it, Tsing found she was "forced to pick out separate, simultaneous melodies and to listen for the moments of harmony and dissonance they created together".[63] The intricate, intertwined lines of the music made her more conscious of the nuances and complexities of the work of listening. She suggests that this kind of attention is needed for responsive and responsible participation within more-than-human assemblages. It is not brought to the text from outside by a reader who employs the same skill with each reading; rather, it is entwined within the material semiotic event. This kind of attention is a constantly attuning inter-folded component of the interaction. It is involved, experimental, risky, multiple, and demanding. These are the characteristics of the kind of attention that Rose describes as practised by her friend and teacher Jessie Wirrpa: "The world was always talking about itself, and Jessie was a skilled listener and observer".[64] Not only was Jessie attentive to the messages of the country, Rose explains that she was also aware of being constantly and carefully attended to by the land. "Jessie Wirrpa lived an ethic of intersubjective attention in a sentient world where life happens because living things take notice".[65] Jessie's skills of taking notice were highly skilled but also practiced continuously, questioned, and refined. Her attention, like that of the other sentient beings around her, was not "diffuse and undifferentiated". It was aware of and focussed along particular story tracks that "cross-cut and overlap each other" across both country and culture, so it entwined along relational stories.[66] In another text, Rose writes of the importance for her Indigenous friends of "tellers", creatures and things who provide information, such as: "When the march flies bite, the crocodiles are laying their eggs". There are many tellers, speaking in "multi-faceted creature-languages".[67] Hearing them can require expert and localised knowledge, and attentiveness that complexly involves connections between all the senses, memory, and technologies. Among Yarralin and Lingara people, Rose explains, paying attention is taught throughout life and becomes honed with practice, guidance, and self-reflection.[68]

Rose notes that taking notice is not necessarily undertaken by people, march flies or mangroves in order to make better environments together, nor are march flies concerned to tell humans when to dig for crocodile eggs. Instead, when all are at stake in one another's thriving, it makes sense to notice what is going on.[69] Creatures are attending to each other within an intra-communicative world "and smart creatures take notice".[70] From this way of understanding, ecological systems are held together through writing and reading both across and within multitudes of perspectivally different relational stories, rather than bounded edges.

Opacity and partiality

Some muddy opacity in storying can be used to share around the power within storying practices. Of the knowledge system of her teachers in the Victoria River region, Rose writes: "A significant aspect of the knowledge system based in country and made up of many participants is that no one knows everything". She continues:

> To be on the inside is to know that one's knowledge does not encompass all the others; to know that country exceeds the knowing of any given knower or any given type of knower. In contrast to a human-centric understanding of knowledge that would see a lack of human knowledge as an epistemological gap waiting to be filled, in creature communities knowledge is widely and patchily distributed. There is a multiplicity of perspectives and knowledges, and there is no privileged perspective. Where one person's or species' knowledge stops, someone else's knowledge picks up the story.[71]

Mangrove mud is an opaque world. Nothing sees far. Relative to many other environments, there are many opportunities to hide, while many perceptions are restricted. For many life forms within the mud, interpretation is made within tight constraints with limited information into risky unknowns. But notwithstanding this opacity, within the mud, multitudes are busy responding to the flows of nutrition that the tides bring. The opacity and restrictedness bring more things together unexpectedly, create more surprising interactions, provide more opportunity for nourishment, for eating and being eaten, for health and disease, for finding niches, and having them destroyed. For something small, a larval protozoon, an algal spore, a tiny fish hatchling, mangroves offer nutrition and moist nooks to nestle without being in full-blown flow. At the same time, many predators use the opacity and viscosity to their advantage for hunting. Mick Smith delves into the "self-revealing/self-concealing soil" and finds, as within the mangrove mud, an ecological community comprising "myriad beings which appear to each other in all kinds of ways, as commensal, as mutualistic, as parasite, as prey, as resources, as co-evolved and evolving beings", making the whole "not entirely comprehensible".[72] While transparency brings the privilege of being able to predict and makes creativity faster and more fluid, opacity slows, enables speculation, makes stories unpredictable, complex, diverse, full of niches and places to hide, and full of hidden things that might be found.

Throughout her work, Haraway calls for maintaining intertwining multiplicities of partial relational stories. She argues that it is through bringing together multiple partial stories that situated knowledges are nourished:

> The knowing self is partial in all its guises. Never finished, whole, simply there and original; it is always constructed and stitched together imperfectly, and therefore able to join with another, to see together without claiming to be another. Here is the promise of objectivity.[73]

Partial, fractured tellings, and dispersed or shared responsibility can operate as storying strategies for what Despret might refer to as practising politeness. Rose describes examples of these strategies as she observed them in careful and frequent use among her Indigenous correspondents. She recounts, for example, the typical manner of instruction of one of her respondents, Old Tim, who:

> offered his wisdom [...] in bits and pieces of conversation that challenge the listener to make the connections. It was not meant to be a closed system. Old Tim kept it open to the world by refraining from formalizing it.[74]

Leaving gaps and spaces within the telling and refraining from finishing the story are methods for allowing room for the interpretation to belong to the listener. The story remains alive in the process of being written-and-read.[75]

Rose demonstrates another strategy, distributing shared responsibilities for storying relations, through the wise words of the Indigenous elder, Mr. Johnson. She recalls:

> His words are emblazoned in my mind: "A good neighbour tells on and on, you know, that same rolling". I can feel the stories rolling across the desert, carrying the local into connectivity and into wider contexts. I hear Mr. Johnson articulating an ethics of story – an ethics of coming face-to-face with neighbors.[76]

Rose explains the rolling across the desert refers to the way stories travel from one story-keeper to another along lines of cultural, historical and geographical meaning-making across the country. All the storytellers work together to keep the relations between these lines of meaning rolling on.[77] A neighbour is regarded as good when they support the story's proper unfurling. Knowing the well-told relational story as ceaseless, as a "rolling on and on" means knowing it requires continual effort and attention, and that it is not closed, that it never ends. It carries its people-and-country along, they roll with it, become with it, it plays its part in sustaining all their stories. These stories determine many roles and responsibilities among their human keepers. They allocate different story-keeping roles and tasks to different people at different sites and at different times. Their tellings are multi-layered in terms of their accessibility; everyone can know some ways of telling but other ways of telling are only for those with special status. The storying makes sure no one person or group can control the whole of the story because no one person or group can know the whole of the story. The story only survives and carries the land-and-people well because it distributes roles and care and responsibility widely and cleverly.

Turning back to the estuary and away from an anthropocentric perspective, ecologist and philosopher Andreas Weber provides an example of the process through which evolutionary adaption deepens the complexity and intimacy, the intensity, of interconnectivity over time.[78] He describes how certain crustaceans have developed specialised mandibles to enable them to eat floating algae which have in turn evolved hard tough body armour to avoid being devoured. Well-adapted specialist species among the crustaceans can eat the toughest-armoured algae, but they have become so specialised they cannot eat any other type of algae. Hence, they spare all the other algae and leave niches for numerous other predators to develop their alternative specialisations. Following Weber, the mangrove grows in the intensity and depth of its being because of the increasing intensities of its entanglements.[79] This intensity or depth, Weber proposes, is the biosphere's beauty. This is a physiological, evolutionary scale of storying, where the material evolutionary process employs a storying strategy of dispersed narrative threads whose entanglement is required to sustain each one. Such strategies can thicken relations; they promote more intense attunements among more relational lines. By this account of ecological aesthetics based upon the intensities of material attentiveness, the mosquito and I share a resonant cellular beauty, continually retuned in response to the perturbations of the virus.

Relationality has been proposed as an important focus for environmental thinking, often associated with an optimism that considering relations might support a shift to greater responsibility in human activity. Through Part 2 "Monstrous relations: exploring a hermeneutic account of relationality" I have argued that if relationality is to meet the optimistic expectations of its proponents, attention should be paid to the work of keeping the active hermeneutic processes of storying relations sensitive to affect, and to keeping them folded rather than layered (in the terms of my ontological discussions in Part 1). Taking a folded rather than a layered approach to relations means (in the terms of this book) avoiding the rigidifications associate with some forms of genre, avoiding privileged perspectives from rigidifying into hierarchical strata, avoiding endings that suggest non-negotiable resolution, and maintaining the potential for plural perspectives. I have reviewed several strategies that are potentially helpful for this work, including Despret's ideas about politeness, Haraway's situated knowledges, kinship, and companion species, Viveiros de Castro's perspectivalism, and Rose's reflections upon indigenous Australian storying strategies that maintain dispersed ownership and fold stories, country and people together. These strategies emphasise the importance of attending to the tensions between holding relations as meaningful within the broader semiotic material contexts that support that meaning, and maintaining their responsiveness within affective, changing situations. This tension is noted as a site of ethicality.

I recover quickly from a slight, undiagnosed illness, never realising the biological entanglements and cellular labour occurring within my body and across our four bodies interconnected by the virus. Summer returns and the pools of water alongside the walking path begin filling with wrigglers. Dragonflies sweep low over the water.

Notes

1 This chapter title is quoted from Donna Haraway, *Staying with the Trouble: Making Kin in the Chthulucene* (Durham and London: Duke University Press, 2016), 12
2 Serenella Iovino and Serpil Oppermann, "Stories Come to Matter," in *Material Ecocriticism*, eds. Serenella Iovino and Serpil Oppermann (Bloomington and Indianapolis: Indiana University Press, 2014): 1.
3 See, for example, María Puig de la Bellacasa, "'Nothing Comes without Its World': Thinking with Care," *The Sociological Review* 60, no. 2 (2012), https://dx.doi.org/10.1111/j.1467-954X.2012.02070.x
4 Karen Barad, *Meeting the Universe Halfway: Quantum Physics and the Entanglement of Matter and Meaning* (Durham and London: Duke University Press, 2007), 323.
5 Donna Haraway, "Staying with the Trouble for Multispecies Environmental Justice," *Dialogues in Human Geography* 8, no. 1 (2018): 104, https://dx.doi.org/10.1177/2043820617739208
6 Thom van Dooren, Eben Kirksey, and Ursula Münster, "Multispecies Studies," *Environmental Humanities* 8, no. 1 (2016): 15, https://dx.doi.org/10.1215/22011919-3527695
7 Eva Haifa Giraud, *What Comes after Entanglement?: Activism, Antrhopocentrism, and an Ethics of Exclusion* (Durham and London: Duke University Press, 2019), 7.
8 Giraud, *After Entanglement*, 11.
9 Haraway, *Staying with the Trouble: Making Kin in the Chthulucene*, 12.
10 Donna Haraway, *Modest_Witness@Second_Millennium. Femaleman_Meets_Oncomouse: Feminism and Technoscience* (New York: Routledge, 1997), 24.
11 Haraway, *Modest Witness*, 37.
12 Uli Beisel, Ann H. Kelly, and Noémi Tousignant, "Knowing Insects: Hosts, Vectors and Companions of Science," *Science as Culture* 22, no. 1 (2013): 7.
13 Margrit Shildrick, "Chimerism and Immunitas: The Emergence of a Posthumanist Biophilosophy," in *Resisting Biopolitics: Philosophical, Political, and Performative Strategies*,: 95–108, eds. A. Žukauskaitė and S. E. Wilmer (New York: Routledge, 2016): 96. For further examples of these narrative tropes, see also Nik Brown, *Immunitary Life: A Biopolitics of Immunity* (London: Palgrave Macmillan UK, 2019).
14 Ed Cohen, "Self, Not-Self, Not Not-Self but Not Self, or the Knotty Paradoxes of 'Autoimmunity': A Genealogical Rumination," *Parallax* 23, no. 1 (2017): 35, https://dx.doi.org/10.1080/13534645.2016.1261660
15 Cohen, "Self", 31.
16 Brown, *Immunitary Life*, 21.
17 See also Michelle Jamieson, "Allergy and Autoimmunity: Rethinking the Normal and the Pathological," *Parallax* 23, no. 1 (2017): 11, https://dx.doi.org/10.1080/13534645.2016.1261659. Here Jamieson provides an account of the limitations of standard immunological framings.
18 Cohen, "Self", 38.
19 Donna Haraway, "The Promises of Monsters: A Regenerative Politics for Inappropriate/D Others," in *Cultural Studies*,: 295–337,eds. Lawrence Grossberg, Cary Nelson, and Paula A. Treichler (New York: Routledge, 1992): 322–324.

Here she explains that the human immune system comprises 10^{12} cells of a wide variety of specialist types, along with acellular components such as antibodies, that circulate between the components of the immune system and between the immune and nervous and endocrine systems.

20 Haraway, "Promises of Monsters", 322.

21 For a study of this process in relation to the development in humans of immunity to malaria, see Denise L. Doolan, Carlota Dobaño, and J. Kevin Baird, "Acquired Immunity to Malaria," *Clinical Microbiology Reviews* 22, no. 1 (2009), https://dx.doi.org/10.1128/CMR.00025-08

22 Haraway, "Promises of Monsters", 323.

23 Ibid.

24 Brown, *Immunitary Life*, 222.

25 Haraway, "Promises of Monsters", 324.

26 Kelly, Ann H. and Javier Lezaun, "Urban Mosquitoes, Situational Publics, and the Pursuit of Interspecies Separation in Dar Es Salaam," *American Ethnologist* 41, no. 2 (May 2014): 368–83.

27 Kelly and Lezaun, "Urban", 103.

28 Bruno Latour, *Reassembling the Social: An Introduction to Actor Network Theory* (New York: Oxford University Press, 2012), 29.

29 Latour, *Reassembling the Social*, 220.

30 Eduardo Viveiros de Castro, *Cannibal Metaphysics*, trans. Peter Skafish (Minneapolis: Univocal, 2014), Kindle.

31 Viveiros de Castro, *Cannibal Metaphysics*, Introduction, paragraph 18.

32 de Castro, *Cannibal Metaphysics*, chapter 1, paragraph 13.

33 Cohen, "Self".

34 Example of these concerns in Haraway's work can be found in Donna Haraway, "Situated Knowledges: The Science Question in Feminism and the Privilege of Partial Perspective," *Feminist Studie* : 14, 575–599 (1988); Donna Haraway, "A Cyborg Manifesto: Science, Technology, and Socialist-Feminism in the Late Twentieth Century," in Donna Haraway *Simians, Cyborgs and Women: The Reinvention of Nature:* 149–181 (New York: Routledge, 1991); Haraway, *Modest Witness.*

35 Examples of these concerns in Deborah Bird Rose's work can be found in: *Nourishing Terrains: Australian Aboriginal Views of Landscape and Wilderness* (Australian Heritage Commission, 1996); "Taking Notice," *Worldviews: Environment, Culture, Religion* 3 (1999): 97–103; "Judas Work," *Environmental Philosophy* 5, no. 2 (2008).

36 Haraway, "Staying with the Trouble for Multispecies."

37 Haraway, *Staying with the Trouble: Making Kin in the Chthulucene.*

38 Donna Haraway, *When Species Meet.*Minneapolis: University of Minnesota Press, 2008, 18.

39 Haraway, *Species Meet*, 85.

40 Donna Haraway, *The Companion Species Manifesto* (Chicago: Prickly Paradigm Press, 2003), 1.

41 Haraway, *Companion Species*, 12.

42 Ibid., 8, 28, 62. See also: "actors become who they are in the dance of relating, full of the patterns of their sometimes-joined, sometimes-separate heritages both before and lateral to *this* encounter. All the dancers are redone through the patterns they enact", in Haraway, *Species Meet*, 25.

43 Haraway, *Companion Species*, 12.
44 Ibid.
45 Michelle L. Gatton et al., "The Importance of Mosquito Behavioural Adaptations to Malaria Control in Africa," *Evolution* 67, no. 4 (Apr 2013), https://dx.doi.org/10.1111/evo.12063
46 Beisel, Kelly, and Tousignant, "Knowing Insects", 10.
47 Ibid.
48 Uli Beisel, "Jumping Hurdles with Mosquitoes?", *Environment and Planning D: Society and Space* 28, no. 1 (2010), https://dx.doi.org/10.1068/d2706wsf
49 Deborah Bird Rose, *Wild Dog Dreaming: Love and Extinction* (Charlottesville and London: University of Virginia Press, 2011), Kindle.
50 "Polite" here intends Despret's sense of the term – this will be more fully explained in the next section of this chapter.
51 Rose, *Wild Dog*, chapter 12, Connectivity Ethics, paragraph 4.
52 This reflects the sense of "written" that Derrida explores through several texts, as I will explore in detail in the third section of this book: "Impossible Differences".
53 Vinciane Despret, "From Secret Agents to Interagency," *History and Theory* 52 (2013): 38.
54 Vinciane Despret, *What Would Animals Say If We Asked the Right Questions?*, trans. Brett Buchanan (Minneapolis and London: University of Minnesota Press, 2016), chapter 5, paragraph 13, Kindle.
55 Despret, *What Would Animals*, chapter 12, paragraph 7.
56 "Agencements" is a Deleuzian term which emphasises how not only the behaviour and materiality, but also the agency of each participant, is inter-enacted within each entanglement.
57 Mirko Nikolić expresses this idea beautifully when he writes: "the world is 'facing' us, listening and yearning", in "Conclusions: For Posthuman Commonings," *Minoritarian Ecologies: Performance before a More-Than-Human World*, 2017, accessed 30 July 2020, http://minoritarianecologi.es/?page_id=1083
58 See for example, in relation to malaria: Doolan, Dobaño, and Baird, "Acquired Immunity to Malaria."
59 Deborah Bird Rose and Thom van Dooren's "ethics of conviviality" offers a similar rendering of this idea of politeness and making room for others, in "Thinking through the Environment, Unsettling the Humanities," *Environmental Humanities* 1 (2012): 2.
60 Haraway, *Species Meet*, 32.
61 Haraway, "Promises of Monsters", 314.
62 Deborah Bird Rose, "Val Plumwood's Philosophical Animism: Attentive Interactions in the Sentient World," *Environmental Humanities* 3, no. 1 (2013): 107, https://dx.doi.org/10.1215/22011919-3611248
63 Anna Lowenhaupt Tsing, *The Mushroom at the End of the World* (Princeton: Princeton University Press, 2015), 23.
64 Deborah Bird Rose, "Taking Notice," *Worldviews: Environment, Culture, Religion* 3 (1999): 97–98.
65 Rose, "Taking Notice", 99.
66 Ibid.
67 Rose, "Val Plumwood's", 104.
68 "Life on the inside of country includes one's self, and thus one's own experience is communicative, too. The work of a lifetime is a never-completed project of

participatory goodness, in which goodness means paying attention and acting properly", Rose writes, at ibid., 105.

69 Thom van Dooren uses the expression being "at stake in each other" to describe multispecies interdependence, in *Flight Ways: Life and Loss at the Edge of Extinction* (New York: Columbia University Press, 2014), chapter 5, paragraph 29, Kindle.

70 Rose, "Val Plumwood's", 104.

71 Ibid.

72 Mick Smith, "Dis(Appearance): Earth, Ethics and Apparently (In)Significant Others," *Australian Humanities Review*, no. 50 (2011), http://press.anu.edu.au/node/1167

73 Haraway, "Situated Knowledges", 586.

74 Rose, *Wild Dog*, chapter 1, Firestick Wisdom, paragraph 5.

75 Bruno Latour expresses the idea with his motto for sociology: "don't fill in the blanks", in *Reassembling the Social: An Introduction to Actor Network Theory* (New York: Oxford University Press, 2012), 246.

76 Rose, *Wild Dog*, chapter 1, Firestick Wisdom, paragraph 4.

77 Rose refers to these stories as Dreaming stories. She explains her usage of this term in ibid.

78 Andreas Weber, *Matter and Desire: An Erotic Ecology*, trans. Rory Bradley. (White River Junction: Chelsea Green Publishing, 2017).

79 Weber, *Matter*, 10.

References

Barad, Karen. *Meeting the Universe Halfway: Quantum Physics and the Entanglement of Matter and Meaning*. Durham and London: Duke University Press, 2007.

Beisel, Uli. "Jumping Hurdles with Mosquitoes?" *Environment and Planning D: Society and Space* 28, no. 1 (2010): 46–49. https://dx.doi.org/10.1068/d2706wsf

Beisel, Uli, Ann H. Kelly, and Noémi Tousignant. "Knowing Insects: Hosts, Vectors and Companions of Science." *Science as Culture* 22, no. 1 (2013): 1–15. https://dx.doi.org/10.1080/09505431.2013.776367

Brown, Nik. *Immunitary Life: A Biopolitics of Immunity*. London: Palgrave Macmillan UK, 2019. https://doi.org/10.1057/978-1-137-55247-1

Ed., Cohen. "Self, Not-Self, Not Not-Self but Not Self, or the Knotty Paradoxes of 'Autoimmunity': A Genealogical Rumination." *Parallax* 23, no. 1 (2017): 28–45. https://dx.doi.org/10.1080/13534645.2016.1261660

Despret, Vinciane. "From Secret Agents to Interagency." *History and Theory* 52 (2013): 29–44.

———. *What Would Animals Say If We Asked the Right Questions?* Minneapolis and London: University of Minnesota Press, 2016, Kindle. Originally published as *Que Diraient les Animaux, si…on leur Posait les Bonnes Questions?*, Éditions La Découverte, 2012.

Doolan, Denise L., Carlota Dobaño, and J. Kevin Baird. "Acquired Immunity to Malaria." *Clinical Microbiology Reviews* 22, no. 1 (2009): 13–36. https://dx.doi.org/10.1128/CMR.00025-08

Gatton, Michelle L., Nakul Chitnis, Thomas Churcher, Martin J. Donnelly, et al. "The Importance of Mosquito Behavioural Adaptations to Malaria Control in Africa." *Evolution* 67, no. 4 (Apr 2013): 1218–30. https://dx.doi.org/10.1111/evo.12063
Giraud, Eva Haifa. *What Comes after Entanglement?: Activism, Anthropocentrism, and an Ethics of Exclusion*. Durham and London: Duke University Press, 2019, Kindle.
Haraway, Donna. *The Companion Species Manifesto*. Chicago: Prickly Paradigm Press, 2003.
———. "A Cyborg Manifesto: Science, Technology, and Socialist-Feminism in the Late Twentieth Century." In *Simians, Cyborgs and Women: The Reinvention of Nature*, edited by Donna Haraway, 149–181. New York: Routledge, 1991.
———. *Modest_Witness@Second_Millennium. Femaleman_Meets_Oncomouse: Feminism and Technoscience*. New York: Routledge, 1997.
———. "The Promises of Monsters: A Regenerative Politics for Inappropriate/D Others." In *Cultural Studies*, edited by Lawrence Grossberg, Cary Nelson, and Paula A. Treichler, 295–337. New York: Routledge, 1992.
———. "Situated Knowledges: The Science Question in Feminism and the Privilege of Partial Perspective." *Feminist Studies* 14 (1988): 575–599. https://dx.doi.org/10.2307/3178066
———. "Staying with the Trouble for Multispecies Environmental Justice." *Dialogues in Human Geography* 8, no. 1 (2018): 102–05. https://dx.doi.org/10.1177/2043820617739208
———. *Staying with the Trouble: Making Kin in the Chthulucene*. Durham and London: Duke University Press, 2016.
———. *When Species Meet*. Minneapolis: University of Minnesota Press, 2008.
Iovino, Serenella and Serpil Oppermann. "Stories Come to Matter." In *Material Ecocriticism*, edited by Serenella Iovino and Serpil Oppermann, 1–18. Bloomington and Indianapolis: Indiana University Press, 2014.
Jamieson, Michelle. "Allergy and Autoimmunity: Rethinking the Normal and the Pathological." *Parallax* 23, no. 1 (2017): 11–27. https://dx.doi.org/10.1080/13534645.2016.1261659
Kelly, Ann H. and Javier Lezaun. "Urban Mosquitoes, Situational Publics, and the Pursuit of Interspecies Separation in Dar Es Salaam." *American Ethnologist* 41, no. 2 (May 2014): 368–83. https://dx.doi.org/10.1111/amet.12081
Latour, Bruno. *Reassembling the Social: An Introduction to Actor Network Theory*. New York: Oxford University Press, 2012.
Nikolić, Mirko. "Conclusions: For Posthuman Commonings." *Minoritarian Ecologies: Performance before a More-Than-Human World*, 2017. Accessed 30 July 2020. http://minoritarianecologi.es/?page_id=1083
Puig de la Bellacasa, María. "'Nothing Comes without Its World': Thinking with Care." *The Sociological Review* 60, no. 2 (2012): 197–216. https://dx.doi.org/10.1111/j.1467-954X.2012.02070.x
Rose, Deborah Bird. "Judas Work." *Environmental Philosophy* 5, no. 2 (2008). https://dx.doi.org/10.5840/envirophil200741/217
———. *Nourishing Terrains: Australian Aboriginal Views of Landscape and Wilderness*. Australian Heritage Commission, 1996. Canberra.
———. "Taking Notice." *Worldviews: Environment, Culture, Religion* 3 (1999): 97–103.

———. "Val Plumwood's Philosophical Animism: Attentive Interactions in the Sentient World." *Environmental Humanities* 3, no. 1 (2013). https://dx.doi.org/10.1215/22011919-3611248

———. *Wild Dog Dreaming: Love and Extinction*. Charlottesville and London: University of Virginia Press, 2011, Kindle.

Rose, Deborah Bird and Thom Van Dooren. "Thinking through the Environment, Unsettling the Humanities." *Environmental Humanities* 1 (2012): 1–5.

Shildrick, Margrit. "Chimerism and Immunitas: The Emergence of a Posthumanist Biophilosophy." In *Resisting Biopolitics: Philosophical, Political, and Performative Strategies*, edited by A. Žukauskaitė and S. E. Wilmer, 95–108. New York: Routledge, 2016.

Smith, Mick. "Dis(Appearance): Earth, Ethics and Apparently (In)Significant Others." *Australian Humanities Review*, no. 50 (2011): 23–44. http://press.anu.edu.au/node/1167

Tsing, Anna Lowenhaupt. *The Mushroom at the End of the World*. Princeton: Princeton University Press, 2015, Kindle.

van Dooren, Thom. *Flight Ways: Life and Loss at the Edge of Extinction*. New York: Columbia University Press, 2014, Kindle.

van Dooren, Thom, Eben Kirksey, and Ursula Münster. "Multispecies Studies." *Environmental Humanities* 8, no. 1 (2016): 1–23. https://dx.doi.org/10.1215/22011919-3527695

Viveiros de Castro, Eduardo. *Cannibal Metaphysics*. Translated by Peter Skafish. Minneapolis: Univocal, 2014, Kindle. Originally published as *Métaphysiques Cannibales*, Presses Universitaires de France, 2009.

Weber, Andreas. *Matter and Desire: An Erotic Ecology*. Translated by Rory Bradley. White River Junction: Chelsea Green Publishing, 2017, Kindle. Originally published as *Lebendigkeit: Eine Erotische Ökologie*, Kösel Verlag, München, 2014.

Part 3

Impossible differences

A muddy journey across more-than-human walls and hospitalities

The next three chapters explore conceptualisations of difference that help to understand the grip of anthropocentrism and human exceptionalism and explore alternatives that are informed through mangrove ecology. My premises are that difference is generated within semiotic material processes and that how this occurs matters. Jacques Derrida's work with différance, the impossible necessity of decision, and democracy to come; Michel Serres' with the parasite; and violence, and hospitality, (topics both examine) are the main concepts considered through this discussion. The chapters follow the story of the dams and embankments of the Cooks River to demonstrate how rigidifying difference establishes problematics of sovereign security and to consider the attractions and violences of making such moves. The later dismantling and reimagining of these Cooks River walls leads to a discussion of approaches that complicate sovereignty. Mangal ways of providing walls introduce alternative approaches to difference aside from the imposition and maintenance of sovereignty by violence.

DOI: 10.4324/9781003286493-10

7 Why build a wall?

Walls make difference become geography, writing it into a landscape. They make a difference to the way the world can be moved through and occupied, to the way it can be used, and they can have this impact upon beings regardless of whether those beings can, or wish to, read the human historical, social, and political significations embodied within the wall. Walls find their ways into maps, language, and ways of thinking, and into concepts of place, body, law, economy, responsibility, and security. They work thought, language, and landscape to assert territory, status, and power. They construct insides and outsides and then gendered and other roles, processes, and meanings become located on one side or the other. In constructing opportunities for rigid inclusion and exclusion, walls turn the values upon which humans base decision-making towards questions of security and identity, and away from questions of interdependent relatedness. Their radical boundary-staking enables human judgements to focus upon interactions within human society, and to turn away from interconnectedness with what becomes construed as outside.

This chapter begins an inquiry into that difference written by the dam and embankment walls of the Cooks River. It traces the colonial history of these walls and draws from that the stories of the power and violence the walls embody. In this chapter, I am interested in understanding the drivers for building these walls, not from a psychological, economic, or political perspective, nor even historically or culturally, but as commitments to certain differences, and to certain storyings of what these differences mean. I am interested in the walls as stakes in these particular differences; in why these differences were, for the people who built those walls, worth such violence and destruction. I am interested in this question because it bears upon the violent stakes humans invest in storyings of difference from other humans, and from other living and non-living things.

A semiotic material wall

The word "wall" signifies something that performs as a hard-to-permeate barrier for something at some time (for water, wind, people, or mud, for

DOI: 10.4324/9781003286493-11

example). Approaches informed by material semiotics recognise that a wall's material meaningfulness as a barrier is a relation that emerges from numerous entangled material narratives. This is not only the case for the wall's material becoming, its being built, but also for its continuing to signify. I may seek a wall motivated by the hope of excluding something or including something but if I employ a builder, she will likely have another motivation; our two motivations co-orient, and their power (whose entangled material semiotic stories involving capitalist economy, engineering, education and more are bracketed here but not through the chapters ahead) constrains and obliges sand, clay, and lime each from their own dispersed contexts into particular co-orientations. A wall becomes and is held in place by the actions of all those, human, mineral, technological, or biological, who, in their co-oriented actions, make it and maintain it, acting from their various historical motivations and constraints. Further, as Tim Ingold points out, "the forms people build, whether in the imagination or on the ground, arise within the current of their involved activity, in the specific relational contexts of their practical engagement with their surroundings".[1] The activities of wanting, planning, and building are affective and relational as much as they are efficacious. A wall becomes and continues, also, because of the stories of all those who are constrained by it, as well as the stories of constraint imagined by those who desired the wall. These unfolding stories of experienced or imagined constraint are enfolded within the semiotic material meaningfulness of that wall along with those of the builders. Paving and roads, a mangrove whelk's shell and a cell wall, a mountain range, or a band of mangroves may count as walls. The meaning is relative to both a specific material and a semiotic context, so it exists relative to something that can read the constraining affect. I can read the whelk's shell as a wall, as can the crab that does not bother trying to eat it and hunts a smaller gastropod instead, but the snail itself may not, and the heron may not as it deftly cracks open the shell. For the animals and microorganisms that quickly set to work eating and rotting them, the original tea-tree wood embankments of the Cooks were food, not an impermeable barrier, so for them these were not a wall. In the chapters ahead, I am interested in the changes effected and prevented by a semiotic material wall; what the river embankments do in the world as they are dreamt, built, dwelt within, understood, written into stories, measured, and valued; as they block, force movement around, enable, protect, deny, and prevent.

Dam: impossibility and difference

Pelican landing

High from the east, from the coast, two pelicans fly slowly, heavily, upriver. As they begin to slowly spiral downwards, gradually lowering themselves, loop after loop, their perspectives shrink from navigating the broad coast and catchment landscape to searching for a landing space for their large bodies along the valley. One comes down low over the river, lowers its legs, and lands into the current, gliding across the water for 200 meters before settling wings and turning into a gracious pelican paddle. The other makes a tighter turn to land more steeply and awkwardly onto a small grassy bank edged by metal embankments, a small harbour dock jutting out into the river near the old sugar factory. Local river watchers successfully lobbied to have a fence erected across one end of this jutting bank, having noticed how favoured it was as a runway and resting place for pelicans and other waterbirds, hoping to protect them from disturbance from dogs and people. Watching them, I am taken back to my own landing, only a few days earlier, into the airport at the other end of this river, my plane circling and lowering, my landscape perspective changing into a local one – the reality of the runway, the exit ramp, the airport sounds and smells. Like the pelican, I was landing onto land reclaimed from the once swampy Cooks River.

Intense disruption to the once extensive Cooks River mangroves began soon after the Sydney settlement as the mangrove trees began to be harvested for use as ash in soap-making and the surrounding banksia and eucalyptus forests cleared for timber and farmland. To provide fresh drinking water and road access across the river, a dam was built across the Cooks River in 1839–40, using convict labour, in the area near the current Princes Highway crossing, just north of the Alexandra Canal. It dramatically changed the water flow, largely cutting off the salty tidal flows and killing all the mangroves upstream, which possibly extended for several kilometres.[2] By the 1880s, the river channel below the dam had already become blocked and inaccessible to larger boats, and substantial dredging and installation of river embankments began. The dredging and embankments facilitated access for boats which were removing timber from the area and cockle and oyster shells for lime production. These were reputedly taken from the large Aboriginal middens along the river, although this is contested.[3] An 800-acre mangrove forest that bordered the river in this area was laboriously removed and the mud flats filled with dredge material. Huge quantities of tea-tree harvested from nearby coastal forests was bound into clumps and used to build embankments along the newly engineered banks,[4] and the reclaimed land became available for cultivation. In 1895, dredging and filling land behind embankments continued upriver from the dam, which had been collecting highly toxic industrial pollution, sewage, and silt to the point of becoming a serious health hazard.[5]

Year by year, through the nineteenth and twentieth centuries, the muddy and fertile in-betweenness of this broad swampy area was replaced by housing, industry, airport, and golf courses. The water was constrained into a channel for freight and a drain for industrial waste and sewage. The men who planned and paid for these walls imagined them as infrastructure for a glorious future whose narrative began in Britain's past and continued colonial social and economic aspirations. These walls would capture and direct a large proportion of the water of the region towards the colonial project and away from all other uses and futures. The dreams of the developers, however, were not fulfilled as imagined. The dam quickly filled with polluted brackish water, the channels were too shallow for freight, and for decades the low-lying housing was flood prone. The upper part of the dam wall and the sluice gates were removed in 1962 with the underpinning parts of the wall finally removed in 1967.[6] Once the dam was removed, the grey mangrove, *Avicennia marina*,[7] found ways back in, colonising upstream as far as Canterbury, and particularly along Wolli Creek, the main river tributary and least altered tidal area of the river, where they now grow thickly.[8] Since the 1970s, human co-conspirators, including the active local group, the "Mudcrabs", have increasingly been working with mangroves and the other native riverbank ecosystems to monitor and support their managed return to the Cooks River.

Although I have argued material semiotics extends beyond entanglements between human language and materiality, the relations between names and places are nonetheless important semiotic materially. The material history of the river is entangled with its history within human languages. How the river and its swamps and mangroves are spoken of, marked on maps, written about in reports and histories continues to matter on the ground and in catchment relations.[9] The Dharug words used by the Kameygal, Bediagal, and Cadigal people to name the wide mangal, mudflats and swampy lowlands of this river valley have not survived the dispossession, dislocation, and cultural eradication that followed Captain James Cook's visit in 1770.[10] The colonial dispossessors were quick to assign a new name: Cooks, a name honouring the English Captain in a colonial practice that covered the land with the names of the conquerors. Increasingly, the arrogance and disrespect of the imposition of that name is being felt in its usage. A new name with close connections to the river is now being adopted by schools, councils, and community groups. The word has been passed down through the generations among Dharawul speakers, the language spoken further down the coast, as the name of a story located in the area, a story about a large and impressive bird that frequently lands in this river estuary. The river is now becoming known as *Goolay'yari*: pelican landing.

Impossibility and différance

The recent history of the mangroves on the Cooks River suggests that what many humans desire in place of mangroves is a wall, a cleared waterway, and no more mangroves, no more of anything but water on one side, land on the other, and a wall in-between. Embankments establish an order that can be mapped, valorised, owned, traded, farmed, built on, or planted with lawn. Along much of the Cooks River, as along many of the coasts and estuaries of tropical and subtropical regions, much human activity is still working hard to create and sustain environments that can be managed and understood within an entirely human-focused framework and that are clean of the otherness of mangroves. Although walls cannot be absolutely impermeable, their violence can hold back the competition (the alternative worldings, those of the mangroves, or of the Kameygal, Bediagal, and Cadigal peoples) long enough for one narrative, (in this case the narrative of British colonial mastery), to inscribe the world with its words and its walls, to turn itself into geography, and position its gatekeepers of categories as sovereign hosts – at least for a while.

In the impossibility of delivering upon their promises, the embankments and dam wall of the Cooks River express a tension, enfolded within the entangled stories composing them, between desires for closure or arrestedness and contrary desires. For Jacques Derrida the tension of impossibility is immanent within the requirements that occur as time and space interact with matter and meaning. Time and space are present only as iterations of past schemas, past arrangements of relationships, that are writing continually into the future, unfolding from inside, but pre-emptively recognisable only because the past schemas retain a currency. The meaningfulness of matter and concept therefore slip into an impossible space between the traces of past schemas refocused, rewritten, into the new context. "An event or an invention is possible only as impossible. That is, nowhere as such, the phenomenological or ontological 'as such' annulling this experience of an impossible that never appears or announces itself as such".[11] "As such" here conveys the sense of the impossibility of anything actually being what the structures or narratives that suggest it – the sylvan river dreams and the stable embankment, for example – promise it to be. Many walls successfully perform their boundary-staking work; their solidity and effectiveness mean they seem profoundly possible. The impossibility that Derrida describes, however, is an impossibility that is folded into the success of a wall.

By dredging, building embankments, and filling in the banks with dredged material, muddy diversity was replaced by clearly defined and simplified categories of river and land. Valuing the assertions of clear category that walls write upon the landscape is the geographical equivalent of the valuing of boundary marking as positive presence that provided the motivation for Derrida's extensive critiques of rationalism and metaphysics.[12] Derrida finds metaphysical accounts of being privilege the positive, the present, and the apparent and as a result consign that which is not complete or present – the muddy, the complex – into the shadows, cast as improper or imperfect others. The negative, the absent, and the in-between are not considered in their own terms because all the terms are filled by metaphysics' basis in the being present of a thing. To redress this Derrida proposes terms of non-presence, including différance and trace, terms that describe erased-things that elude presence but are nonetheless necessary within every sign and interpretation. Différance, the differing and deferring that writes the world, represents the possibility of a way of knowing and speaking beyond concept, word, sign, or semiotic system.[13] It is something that precedes these and is necessary for them; an originary shift, the becoming of a difference. It describes the unspeakable non-concept that nonetheless occurs; and once it has occurred, patterning, folding, iteration, the organisations of semiosis, of potential meaningfulness figured among, can follow. "Différance is the non-full, non-simple, structured and differentiating origin of differences".[14] Reaching any stable meaning is impossible because there is always différance within any meaning-finding. To my reading, Derrida's work with (non)concepts such as différance and impossibility continually explores questions of meaningfulness and difference within a semiotic materiality.[15]

Parasite

In his notion of the parasite, Michel Serres approaches impossibility in a way that contrasts helpfully with Derrida's, while sharing a similar close focus upon the site of difference and finding a similar sense of tragic excitement.[16] The parasite as Serres conjures it may be an organism nourishing itself from the bodily fluids of another organism, but it may also be a noise, a guest in the house of a host, or an arrow intent on its own directionality. In short, it is whatever interferes, irritates, or intrudes, purely to satisfy its own interests, to demand a change to the would-be continuity of the directionality of another thing or process. The parasite is the irritation that demands response. That demand is the source of difference, change, or divergence, of adaptation or re-equilibration. It means the jitteriness of life and of a lively universe as directionalities interfere with and respond to each other in all manner of ways, some destructive, some sustaining, some generative:

> The parasite is an exciter. Far from transforming a system, changing its nature, its form, its elements, its relations and its pathways [...] the parasite makes it change states differentially. It inclines it. It makes the equilibrium of the energetic distribution fluctuate.[17]

The parasite is relevant beyond living systems: any prescient directionality that begins to form will be irritated by a parasite. Serres writes of earth as parasite of sun's energy, of wind as parasite of wall: the same arrow, the same kind of interference across all relationships. Like Derrida, Serres finds difference fundamental to being: "the difference is part of the thing itself, and perhaps it even produces the thing. Maybe the radical origin of things is really that difference".[18] If not for difference, there would be nothing to notice. This difference involves the sense of impossibility at every turn, as it does for Derrida. But impossibility for Serres is not about impasse or aporia, however miniscule or fleeting, in there at the heart of things, as it is for Derrida. For Serres, impossibility involves the frustration within every diverted directionality. Here is an impossibility of stasis, an impossibility of rest, or of settling into sameness. The ceaseless annoyances, the agitations, the bothersome interventions that demand responses make for a noisy, edgy impossibility.

The action of the parasite always occurs within relationship, in the between, "in the most profitable positions, at the intersection of relations".[19] In-between, there is always a third, the parasite: the one that makes difference, that makes time, that keeps closure impossible. Parasites find opportunities between; they fill the spaces there, making difference from inside relations that are already unfolding. The parasite is one of Serres' stories for what relationship is, for how to get in there inside whatever it is that is happening when what we identify as things do what we identify as affecting each other. Although the guest/parasite uses the host, there is something the host has already received – the position of having capacity in the first place, which the host has by virtue of being also a parasite upon some other host. With host, comes immediately, parasite, and parasite quickly becomes host. To get rid of the parasites one would have to destroy the system – the whole world – as Serres suggests René Descartes indeed did,[20] in order to create a new logocentric world with rigid walls to keep out parasites.

Parasitism makes for a strategy-filled world, jittery with experiments in accommodation, resistance, opportunism, generosity, reluctance, caution, boldness, attraction, repulsion. In living systems, the parasite "brings us near the fine equilibria of living systems and near their energetic equilibria. It is their fluctuation, their moving back and forth, their test and training".[21] The parasite acts as the tuning of the negotiations of which materiality is comprised, the continuous irritations that nudge life to adapt. In the mangrove, the negotiations of predation; of mutual nudging into accommodation; of energy distribution, degradation, and combustion; infection and immunity are sorting, jostling, and ordering, not to reduce a system to simplicity or balance, nor to radically transform it, but because of all the parasites.

The violence of a wall

Dredging, mining, impeding, and imposing

Dredging mud from waterways to deepen and straighten channels and filling up low-lying marshy land with that dredged material is widely practiced to change mangrove-edged estuaries into economically valorisable places. The process is enormously violent. Removing mangroves is difficult and along with the trees, multitudes of other living beings are killed or forced away. Breeding areas, rich food sources, and shelter for fish, reptiles, marine mammals, birds, crustaceans, and many plants are destroyed. The walls themselves are created from materials mined or chopped and carted from other locations and dredging the riverbeds to remove the silt and mud destroys the river ecology, killing crustaceans living in the mud and fish that get sucked up with the dredge material, and eradicating seagrass beds, to the river side of the mangroves. Continued dredging is likely to be needed. Ian Tyrrell records that building the Cooks River dam and the later work to remove mangroves and build embankments was "arduous"[22] and "backbreaking" with many workers becoming trapped in mud or falling ill.[23] Not only was a large mangrove forest removed, but also a lush forest of tea-tree fronting a nearby beach, for use as the construction material.

Entire mountains, forests, and beaches have been destroyed and transformed into the timber, bricks, stone, and concrete of walls. Mining sand from tropical rivers and sand dunes, and logging and mining in the catchments of rivers affect areas of mangroves dramatically, as well as other ecosystems. Sand mining for concrete and for land reclamation causes enormous ecological damage.[24] The quantity of sand and gravel mined globally exceeds that of fossil fuels, comprising approximately 68–85% of the material mined globally.[25] Pascal Peduzzia estimates the amount mined to be more than twice the sediment load of all the world's rivers combined.[26] Sand provides a habitat for a wide range of animal species; ribbon worms, molluscs such as clams and snails, arthropods such as crustaceans, spiders, and insects, and echinoderms such as sand dollars. Additionally, huge numbers of tiny microfauna live between the sand particles.[27] All these living things who dwell in the sand are killed during sand mining, ending all the potential future generations of their offspring. Many of the species of turtles, penguins, and shorebirds who lay eggs in or on sandy beaches are now facing extinction. Although sand mining and the building of sea walls are not the sole causes of this, they are contributing factors towards reducing, degrading, or making inaccessible the habitat these animals require for laying eggs. Most of those who lived in the forests harvested or the mountains mined to build walls were similarly killed or driven away to face perilous futures.

Furthermore, the physical presence of walls, fences, and the horizontal walls that are roads and paving, violently impedes the movement and fragments the living space of non-humans. Because walls demand that non-humans (those who have not specifically been invited to "come inside" in the process known as domestication, whose movement is arguably even more narrowly constrained) have to move and live around them, movement is continually impeded, and greater effort and risk is constantly required to access water, food, and kin. Walls force non-humans and many humans to become marginal in relation to worlds where the combined vision, power, and insecurity of certain humans have constructed the geography of possibility. In addition to all this, walls require ongoing violence for their maintenance, renovation, and expansion. Aside from the physical brutality, walls are violent in more conceptual ways, ways which Derrida and Serres have introduced and elaborated.

Walls demonstrate the three types of violence that Derrida discusses: originary violence, the violence of iterative maintenance, and the violence of everyday conflict.[28] Originary violence (the violence of inscription within a system, the "pure performative act that does not have to answer to or before anyone"[29]) is well exemplified by the violence of initially building a wall. This is the act of violence through which an authority (within the meaning schema of a given semiotic) imposes its power. From my semiotic materialist approach, this originary violence is perpetrated by one storying of the landscape upon the innumerable other stories of territory that previously wrote the landscape into functional meaning for so many others, human and non-human.[30] We build the Anthropocene – which is the semiotic material interpretation of the world as anthropocentric – from the destruction of the semiotic material worlds of other world makers (including non-anthropocentric humans, non-humans, and more-than-human collectives).

The secondary, protective, violence Derrida describes is that which is required to maintain the authorities' power, generally law-making and the upholding of law and conventions through a wide range of mechanisms including economic mechanisms, social norms, and enforcement. It is "a second violence that is reparatory, protective, instituting the 'moral' ".[31] In the case of walls, property laws and economic status are required to make the walls meaningful and maintainable. Even as the initial founding violence occurs, it is, Derrida claims, engaging in an iteration of itself, as the founding act signals already its requirement for continued preservation. Building a wall not only signals having the capacity to build it, but also the capacity to maintain it. There is a violent and energetic conservative optimism built into the wall, the original violence founds "a promise" to "heritage and to tradition".[32] Derrida's third violence occurs when these first two violences are challenged through, for example, in the case of walls, trespass, vandalism, or mangrove roots damaging the embankments. This violence poses a threat to the "secret" of the first two violences by suggesting, in the case of walls, that walls are not in fact impermeable.[33] This is the violence to which the word "violence" will be applied within the value context established by the first two.

Existing walls, if we apply Derrida's thoughts about foundational and secondary violence, actively continue reimposing the violence of their original foundations. They re-enact the power relationships of their imposition and as such are conservative forces within and upon social, economic, political, legal, and biological landscapes. Originary violence is most successful when it is forgotten, when the original violence re-membered within the stone and wood and legal structures that constrain and direct decisions, actions, narratives, and ambitions becomes normalised as the stable way of things: "successful unifications or foundations only ever succeed in making one forget that there never was a natural unity or a prior foundation".[34] For there to be a wall, there is already repetition, implied within the edifice. The border that the wall proclaims must be meaningful from time to time. A wall is an optimistic pronouncement of the possibility of this repetition. The original enactment, which carries, quite concretely in the case of a wall, the trace of the desires, the tastes, the power dynamics, the technological capacities, the economic milieu of the situation which lead towards its building, is rejuvenated, repeated continually in the ways beings – water, mud, humans, mangroves, cyclones – are affected by the wall. The wall builders desired to assert their territory in order to affect the trajectories of others in just these ways. So long as those trajectories continue to be affected, the will, the narrative thrust behind the territory is repeated upon the landscape. A Derridean deconstructive approach would retrace the motivations behind the walls, look for the clues to what the embankments intend, what the problems were that the walls were constructed as the answer for. This might enable a reimagining of the walls or lead perhaps to discovering these problems are no longer relevant or are answerable using an alternative approach.

Like Derrida, Serres' continually returns to find violence in an originary difference. But where Derrida finds originary violence in the performative act of inscription, Serres finds it in deviation. Everything is always in motion for Serres; trajectory is implicit, originary, and tireless, and violence happens because trajectories buffet and parasitise each other. Each trajectory is moving along like an arrow arc; it is nothing but its direction: "The first economic relation is of abuse. But when the arrow does not kill, when abuse does not pass the point of no return, the relation can evolve toward another equilibrium".[35] The violence of irritation, of abuse, is inevitable where there is change or difference. What matters is how it is engaged with, how attunement can be found. For Serres, unless steps are taken – a contract, a placation, a gift, a wall, for example – interactions are likely to be violent. Both thinkers note that power in the form of originary violence strategically hides itself and becomes invisible. Originary violence is well disguised inside the solid barrier of a wall. The most successful parasite, for Serres, is the one who becomes so well incorporated within the body of its host that the host fails to notice.

Serres writing engages with different violences at different scales and times. There is the intimate violence of the parasitic arrow, the violence of death that is needed for existence, and the violence of the exile required to face death and in so doing become more alive. These three are instances of a generative, provocative violence that while it is disturbing and discomfiting, also provides the irritation that teases out adaptive innovation and functional forms of political process[36] and mutually responsive attentiveness and communications. Counter to these is another violence, instantiated in the violence of the ploughshares and expulsions of agricultural society, the violence of analysis, the violence of cleaning out, the violence that founds territory and empire. This violence is that of the wall, a violence that seeks to demarcate and protect a marked-off inside while excluding a simplified outside:

> The ploughshare is a sacrificial knife frenetically manipulated at the height of murdering fury [...] It marks a closed line: inside, the sacred;

> outside, the profane; inside, the temple; outside, the vague area filled with evil [...] space everywhere is sliced up [...] everything has been worked into a fine powder. Harrowed. Reduced to its elements. Analysis.[37]

Between these two violences, Serres offers his path of negotiation, which he tends to call the contract. Violence can be quieted by a contract, which for Serres is something like an open, informed stance taken together to the mutual value of the relationship. To make this requires work and in particular, attention. This provides a brief equilibrium that we may read as justice: "This rarity is sometimes called justice. A difficult effort, an exceptional, miraculous, human one".[38] Such justice, for Serres, requires some breaking down of protective walls, some preparedness to become excluded, to be exposed, to be vulnerable to parasites.

While they have much in common, the differences between Serres and Derrida in their understanding of originary violence are significant and underscore the basic differences in their semiotic material ontologies. Derrida's difference plunges everything into a trembling stasis of undecidability, held in the suspension of its impossibility, with a duty to leap chaotically into the next state of suspension. Serres' difference, by contrast, forces everything to keep moving constantly, interactions are always affective, demanding that the impossible be continually encountered.

Banks: staking a story

To have, to hold, to except

Being in the position to build a wall implies having already achieved the capacity to do so, and that something has already been achieved that is evaluated as potentially losable. There is a desire to stop or delay time to keep a position of relative security and power so the narrativity of that security can, at the very least, continue. Walls are tools to hold onto a narrative thread in a material semiotics that has already presented itself as potentially worth keeping within some evaluative frame. Behind the desire to build a wall is an imagined future for something that can be held onto and can continue from the past; where some form of sameness will continue and be able to grow or change as is most flourishing for it; where there can be a delay of continual difference, so the same can have the scope, the space, and time to become its best. Matthias Fritsch provides this Derridean description of the emergence of normative patterns within nature, patterns that become and repeat the same into an identity:

> An organismic identity, then, does not come into the world ready-made, or nonrelationally, identical to itself. It must strive for that identity, for its time and its space. With this striving, normativity – the "must" or "il faut" – comes into the world. If the identity of an element, such as

> an organism, is spaced-out, a little mad in not already coinciding with itself, it can be itself only by being identified, again and again (hence the deferral), with itself. Its identity must rely, and from the beginning, on a repetition that intends the same.[39]

The imagination that desires a wall, the creative urge behind all that effort, is one of holding back change, holding back interruption, so as to enable some state of sameness to persist "again and again" and possibly thrive.

The sameness driving the enormous violence of the dam and the first embankments on the river lay in the underlying narratives of the developers. Tyrrell records how they described their plans and visions as narratives of improvement, of advancement towards a glorious future way of life for themselves and their descendants that would be not only aesthetically pleasing and historically significant, but also morally superior.[40] They envisaged a future that would look, feel, and signify very differently from the present as they were experiencing it. These dreamers were a particular group of privileged men whose ideas of improvement were very closely aligned with the values, tastes, and aspirations of the society they sought to establish themselves as worthy within. The "same" they were building their walls to affirm and nourish was the social, moral, and economic order of Victorian industrial England as they imagined it. They were imagining themselves as well-positioned within a globally successful colonial narrative. The narrative they wished to strengthen and protect esteemed the man who was prepared to invest in large-scale engineering for the sake of moral, scientific, and aesthetic improvement, as evaluated by the standards of that society. By contrast, the embankments of the Cooks River brought great change to the land of the Kameygal, Cadigal, Bediagal, and Wangal people of the Eora nation.[41] Like all the walls of colonial imposition, for those already present, they signal overwhelming cultural destruction and breakage rather than continuity. The boundaries of the colonial system and its epistemologies of land violently ruptured indigenous regimes of land tenure and usage which were validated and maintained through complex social practices.[42]

If it has come to be built, a wall is already embodying power; and in being built, it establishes that power as sovereign, as an assertion of a dominion within which the wall-builder has the power to exclude. Exclusions are productive, they make difference to the world and to the future.[43] The material fact of physical inscription upon the landscape forces material compliance. Walls demand that the landscape complies with their social and political significance. Additionally, legal, economic, and social edifices layer walls with meaning such that they are laden with the enabling and enforcing directives and dreams of those who had and have the power to create and maintain them. The messages of the Cooks River walls required interpretation through the edifice of property law, and more fundamentally through the assumption of anthropomorphic exceptionalism. Indeed, any human-built wall indicates that certain humans might take it upon themselves to reshape the world to benefit those who can accumulate the power and resources to build a wall, building that power into the landscape.

Derrida engages with the question of sovereignty, drawing upon Schmitt's conceptualisation of the sovereign as the one who has the right to define the exception, the state in which agreed rules can be disregarded and replaced by emergency provisions. Building a wall defines exception directly and materially, through a violently creative act that forcibly stops the other forms of negotiations of territory. A wall by this account claims territory as emergency; establishing an exception through which, in the case of house walls, the wild and the homeless are defined and excluded, thereby identifying the insider, in this case humans and domesticated non-humans. An embankment separates river from land through mutual exclusions, making each available to the treatments established for it according to the human social and economic norms of the embankment builders. Georgio Agamben, in his investigation of Schmitt's thesis, points out that the exception does not, however, remain entirely on one side of the wall. While the establishing of an exception "opens the space in which the determination of a certain juridical order and a particular territory first becomes possible",[44] sovereignty also requires the exception to be brought across in a particular form. The exception comes across in a scaled-down, abstracted form where, up close and in detail, it can be worried over and analysed, now an object of anxiety. Further, it can become a key to the aesthetic, ethical, and cultural provocations that steer the shaping of social values. In this way, the identity of the resident is coproduced with that of the homeless, who provide the exception that nominates the insider as successful by virtue of not being outside. The fact of homelessness also provides the basis for ethical and political discourse around responsibility for those positioned on the margins, discourse where the boundaries of social inclusion can be refined and debated. Mud, likewise, becomes a worry; the hearth and the laundry the troublesome in-betweens where it is actively engaged with, worried over, transformed into something drier or something wetter, filtered into the included or excluded.

This fascination with and anxiety about the abstracted cross-over exception occurs in the case of the Cooks River embankments. When embankments define the dry world of property development and land ownership from the wet river world, the exception that occurs during heavy rainfall when the wet world impinges over onto the dry becomes the subject of anxiety, and of detailed and specialised attention. The broad in-between of the swampy low-lying valley and its tidal estuary became flood-prone once its boundaries were made and mapped. In 1841, not long after the dam was built across the river at Tempe, floods overflowed its barricade. Subsequent floods through the 1840s washed part of the wall away.[45] This was the beginning of ongoing and costly work to repair and redesign the dam and the water flow across the catchment, work which "inflamed tensions within communities, and between one suburb and another".[46] Real-estate subdivision continued on flood-prone land as the city expanded, as did the remedial work to drain and manage the situation, and the significant suffering of the residents. A pumping station and retention pond began controlling the floodwaters mechanically from 1935. Where dams and embankments delineate a difference between wet and dry, floods are a challenging exception. I will consider the role of the Cooks River mangroves as challenging exceptions to the colonial redefinition of the Sydney riverine landscape, as well as their relationship with the river's floods, in Chapter 9.

Notes

1 Tim Ingold, *The Perception of the Environment: Essays in Livelihood, Dwelling and Skill* (London and New York: Routledge, 2000), 186.
2 Ian Tyrrell, *River Dreams: The People and Landscape of the Cooks River* (Sydney: University of New South Wales Press, 2018), chapter 4, paragraph 9, Kindle.
3 Paul Irish, *Aboriginal History Along the Cooks River* (Sydney: Cooks River Alliance, 2017), 27.
4 Tyrrell, *River*, chapter 5, The Aesthetics of Dredging, paragraph 4.
5 Vanessa Witton, "Damming the Cooks River," *The Dictionary of Sydney, City of Sydney and State Library NSW*, 2013, accessed 1 March, 2018, https://dictionaryofsydney.org/entry/damming_the_cooks_river
6 Tyrrell, *River*, chapter 7, "Valley of Industry valley of pollution", paragraph 17.
7 It seems likely that the other previously common mangrove tree species, *Aegiceras corniculatum,* the red mangrove, has not yet returned.
8 K. R. Brown et al., "The Mangroves of Wolli Creek, Sydney," *Wetlands Australia Journal* 8, no. 1 (1988): 27–29.
9 For a discussion of the importance of the semiotic contrasts between Indigenous and colonial settler Australian language presumptions, particularly regarding relationality, see Helen Watson and David Wade Chambers, *Singing the Land, Signing the Land* (Geelong: Deakin University Press, 1989).
10 Tyrrell, *River*, chapter 1, note 12; Irish, *Aboriginal*, note 18.
11 Jacques Derrida, *Rogues: Two Essays on Reason*, trans. Pascal-Anne Brault and Michael Naas (Stanford: Stanford University Press, 2005), 144.
12 Jacques Derrida, *Of Grammatology*, trans. Gayatri Chakravorty Spivak (Baltimore: Johns Hopkins University Press, 2016), 13, Kindle.
13 Jacques Derrida, *Margins of Philosophy*, trans. Alan Bass (Brighton: The Harvester Press, 1982), 9.
14 Derrida, *Margins*, 11.
15 Derrida states repeatedly that he uses linguistic terms such as "writing" or "text" beyond a human context, within a broad semiotic materialism. A sustained account of Derrida's non-anthropocentric position is made in Vicki Kirby, "Grammatology: A Vital Science," *Derrida Today* 9, no. 1 (2016), https://dx.doi.org/10.3366/drt.2016.0119
16 Andrew Gibson is perceptive on the richness and density of Serres' conjuring of impossibility or aporia: "For the bleak austerity of aporia, then – the thought that a parting or crossing of ways means less and not more, loss

rather than gain – Serres substitutes the thought of the juncture as abundance or complexification", in "Serres at the Crossroads," in *Mapping Michel Serres*, ed. Niran Abbas (Ann Arbor: The University of Michigan Press, 2005), paragraph 11, Kindle.

17 Michel Serres, *The Parasite*, trans. Lawrence Schehr, (Minneapolis: University of Minnesota Press, 2013), The Best Definition, paragraph 6, Kindle.
18 Serres, *Parasite*, Rats' Meals Cascades, paragraph 36.
19 Ibid., Pentecost, paragraph 8.
20 Ibid., Rats' Meals Cascades, paragraph 34.
21 Ibid., The Best Definition, paragraph 8.
22 Tyrrell, *River*, chapter 3, paragraph 4.
23 Ibid., chapter 5, paragraph 7.
24 A documentary exploration of this destruction is presented in Kalyanee Mam, "Lost World," *Emergence Magazine*, 2020, https://emergencemagazine.org/story/lost-world/. Mam follows the destruction of the mangroves of Cambodia's coast for sand to build the land reclamation of Singapore. She notes "Since 2007, Singapore imported over 80 million tons of sand", from the mangroves of Cambodia.
25 Pascal Peduzzia, *Sand, Rarer Than One Thinks* (Coastal Care: UNEP Global Environmental Alert Service, 2014), https://na.unep.net/geas/getUNEPPageWithArticleIDScript.php?article_id=110
26 Peduzzia, *Sand.*
27 Thomas Schlacher et al., "Open-Coast Sandy Beaches and Coastal Dunes," in *Coastal Conservation*, ed. Brook Maslo and Julie Lockwood (Cambridge and New York: Cambridge University Press, 2014).
28 Derrida, *Grammatology*; Jacques Derrida, "Force of Law: The Metaphysical Foundation of Authority," in *Deconstruction and the Possibility of Justice*, ed. Drucilla Cornell, Michel Rosenfeld, and David Carlson (New York and London: Routledge, 1992).
29 Derrida, "Force", 36.
30 Dinesh Wadiwel, discussing Derrida's position on violence, finds it recognises "humans are not innately superior to animals, but win their superiority through the application of force", in "The Will for Self-Preservation: Locke and Derrida on Domination, Property and Animals," *SubStance* 43, no. 2 (2014): 149, 155, https://dx.doi.org/https://doi.org/10.1353/sub.2014.0029
31 Derrida, *Grammatology*, 121.
32 Derrida, "Force", 35.
33 Derrida, *Grammatology*, 121.
34 Derrida, cited in Nick Vaughan-Williams, "Borders, Territory, Law," *International Political Sociology* 2 (2008): 327.
35 Serres, *Parasite*, "Borders", Meals of Chestnuts, paragraph 11.
36 Particularly relevant here is Michel Serres, *The Natural Contract*, trans. Elizabeth MacArthur and William Paulson (Ann Arbor: The University of Michigan Press, 1995). Here Serres argues for the inclusion of nature into contract with humans as a partner in political decision-making.
37 Serres, *Parasite*, The Cows come out of the River, paragraph 11.
38 Ibid.
39 Matthias Fritsch, "An Eco-Deconstructive Account of the Emergence of Normativity in 'Nature'," in *Eco-Deconstruction: Derrida and Environmental*

Philosophy, ed. Matthias Fritsch, Philippe Lynes, and David Wood, (New York: Fordham University Press, 2018), Derrida's Double Affirmation, paragraph 5, Kindle.

40 Tyrrell, *River*, Introduction, paragraph 6.

41 Anita Heiss and Melodie-Jane Gibson, "Aboriginal People and Place," Barani: Sydney's Aboriginal History, last modified 2013, accessed 12 March 2020, www.sydneybarani.com.au/sites/aboriginal-people-and-place

42 Watson and Chambers, *Singing the Land, Signing the Land*, 44. Watson and Chambers describe how land titles in the form of pieces of paper would now be the treasured and securely protected arbiters of land allocation and control. Land title maps work more efficiently when their lines are clear and thin, like the river's new embankments.

43 Eva Haifa Giraud suggests "there is a need to recognize not just the constitutive role of certain exclusions but their productive role: that purposeful acts of contesting particular relationships are sometimes necessary to create space for alternatives to emerge", in *What Comes after Entanglement?: Activism, Anthropocentrism, and an Ethics of Exclusion* (Durham and London: Duke University Press, 2019), 11. Terrence W. Deacon and Tyrone Cashman take a more radical position, arguing constraints are fundamental to ontology. They argue it is by the reproduction of constraints that things continue, and by the change of constraints that they change, in "Eliminativism, Complexity and Emergence," in *The Routledge Companion to Religion and Science*, eds. James W. Haag, Gregory Peterson, and Michael Spezio (Oxon and New York: Routledge, 2011): 199.

44 Georgio Agamben, *Homo Sacer: Sovereign Power and Bare Life*, trans. Daniel Heller-Roazen (Stanford: Stanford University Press, 1998), chapter 1, paragraph 9, Kindle.

45 Tyrrell, *River*, chapter 4, paragraph 17.

46 Ibid., chapter 4, paragraph 19.

References

Agamben, Georgio. *Homo Sacer: Sovereign Power and Bare Life*. Stanford: Stanford University Press, 1998, Kindle.

Deacon, Terrence W. and Tyrone Cashman. "Eliminativism, Complexity and Emergence." In *The Routledge Companion to Religion and Science*, edited by James W. Haag, Gregory Peterson, and Michael Spezio, 193–205. Oxon and New York : Routledge, 2011.

Derrida, Jacques. "Force of Law: The Metaphysical Foundation of Authority." In *Deconstruction and the Possibility of Justice*, edited by Drucilla Cornell, Michel Rosenfeld, and David Carlson, 3–67. New York and London: Routledge, 1992.

———. *Margins of Philosophy*. Translated by Alan Bass. Brighton: The Harvester Press, 1982. Originally published as *Marges de la Philosophie*, Les Éditions de Minuit, 1972.

———. *Of Grammatology*. Translated by Gayatri Chakravorty Spivak. Baltimore: Johns Hopkins University Press, 2016, Kindle. Originally published as *De la Grammatologie,* Les Éditions de Minuit, 1967.

———. *Rogues: Two Essays on Reason*. Stanford: Stanford University Press, 2005. Originally published as *Voyous: Deux Essais sur la Raison,* Éditions Galilée, 2003.

Fritsch, Matthias. "An Eco-Deconstructive Account of the Emergence of Normativity in 'Nature'." In *Eco-Deconstruction: Derrida and Environmental Philosophy*, edited by Matthias Fritsch, Philippe Lynes, and David Wood, 279–302. Groundworks: Ecological Issues in Philosophy and Theology. New York: Fordham University Press, 2018, Kindle.

Gibson, Andrew. "Serres at the Crossroads." In *Mapping Michel Serres*, edited by Niran Abbas, 84–98. Ann Arbor: The University of Michigan Press, 2005.

Giraud, Eva Haifa. *What Comes after Entanglement?: Activism, Anthropocentrism, and an Ethics of Exclusion*. Durham and London: Duke University Press, 2019, Kindle.

Heiss, Anita and Melodie-Jane Gibson. "Aboriginal People and Place." Barani: Sydney's Aboriginal History. Last modified 2013. Accessed 12 March 2020. www.sydneybarani.com.au/sites/aboriginal-people-and-place

Ingold, Tim. *The Perception of the Environment: Essays in Livelihood, Dwelling and Skill*. London and New York: Routledge, 2000.

Irish, Paul. *Aboriginal History Along the Cooks River*. Sydney: Cooks River Alliance, 2017.

Kirby, Vicki. "Grammatology: A Vital Science." *Derrida Today* 9, no. 1 (2016): 47–67. https://dx.doi.org/10.3366/drt.2016.0119

Mam, Kalyanee. "Lost World." *Emergence Magazine*. 2020. https://emergencemagazine.org/story/lost-world/

Peduzzia, Pascal. *Sand, Rarer Than One Thinks*. Coastal Care: UNEP Global Environmental Alert Service, 2014. https://na.unep.net/geas/getUNEPPageWithArticleIDScript.php?article_id=110

Schlacher, Thomas, Alan Jones, Jennifer Dugan, Michael Weston, et al. "Open-Coast Sandy Beaches and Coastal Dunes." In *Coastal Conservation*, edited by Brook Maslo and Julie Lockwood, 37–94. Cambridge and New York: Cambridge University Press, 2014.

Serres, Michel. *The Natural Contract*. Ann Arbor: The University of Michigan Press, 1995. Originally published as *Le Contract*, Editions François Bourin, 1992.

———. *The Parasite*. Minneapolis: University of Minnesota Press, 2013, Kindle. Originally published as *Le Parasite*, Grasset et Fasquelle, 1980.

Tyrrell, Ian. *River Dreams: The People and Landscape of the Cooks River*. Sydney: University of New South Wales Press, 2018, Kindle.

Vaughan-Williams, Nick. "Borders, Territory, Law." *International Political Sociology* 2 (2008): 322–388.

Wadiwel, Dinesh. "The Will for Self-Preservation: Locke and Derrida on Domination, Property and Animals." *SubStance* 43, no. 2 (2014): 148–61. https://doi.org/10.1353/sub.2014.0029

Watson, Helen and David Wade Chambers. *Singing the Land, Signing the Land*. Geelong: Deakin University Press, 1989.

Witton, Vanessa. "Damming the Cooks River." *The Dictionary of Sydney. City of Sydney and State Library NSW*. 2013. Accessed 1 March 2018. https://dictionaryofsydney.org/entry/damming_the_cooks_river

8 Walls and human exceptionalism

It takes a lot of work to remove mangroves and replace them with embankments and sea walls. Then it takes more work to mend, to repair, strengthen, dredge, uproot, poison, to remain vigilant against the unwanted, untidy mangrove intruder. In general, the trouble humans go to in building walls, especially massive sea walls, is stunning. In general, human wall builders invest vast proportions of their time, money, and earth's resources into the work of building walls. Why would we embark upon such work? Why go to so much trouble? The answer is in the scale of the benefits that walls provide. They quieten the world and enable the possibility of rest and of security without constant vigilance. This extends even to making it possible to still the tides, to arrest the changing narratives of tidal worlds. The apparent possibilities of human exceptionalism, of the opportunity to manage and control thought and materiality, and of the establishment of stable identities are facilitated by walls. "For there to be world enough and time, the system must be absolutely closed" writes Michel Serres, a comment he follows immediately by its necessary complicator: "But it is open".[1] This complication will be considered in depth soon, first it is worth exploring here the immense pull towards wall building.

In this chapter, I am interested in how walls have contributed towards human exceptionalism, exemptionalism (the assumption that humans can be exempt from ecological interdependencies), and anthropocentrism.[2] Without walls, territorialising activity involves attentive concern to relational work across the interfaces of responsive interactions. I argue that human walls have made it possible for their builders to justify positioning themselves as winners within relational stories that write territorialising as strategic and winnable. I do not mean to infer that this is always the case. It is possible for human walls to not do this, as I will discuss in the final chapter. The activities of walling and cleaning support each other in a story that aspires towards purity, rest, and stasis, where the violence of achieving these is hidden in the architecture.

DOI: 10.4324/9781003286493-12

Territory

Territorialising stories

Without a wall, territory has to be negotiated somehow. Communication of some kind within relationship is required, some ongoing relationship work. When Stuart Elden, in his account of the genealogy of the Western concept of territory, notes that "territory is itself a process, made and remade, shaped and shaping, active and reactive",[3] he suggests something of this activity. But the processes from which territories emerge are not limited to human activities. Territorialising strategies abound: displays of colour, calls, scents, stories, rituals, regular or irregular bodily movement, diplomatic conversations, the recording of financial transactions, the emission of chemical signals, the maintenance of standing armies, physical spreading of leaves or roots or genetic descendants, aggressive and submissive behaviours are a few that come to mind. These are all semiotic, require interpretation by others, and all can fail. They also require repetition and refinement. Making territory is typically demanding ongoing storying work. For some (male bowerbirds, border guards, lawyers, and immune systems quickly come to mind), territory work is a predominant part of the work of living.[4] Furthermore, it is not the case that asserting territory ceases beyond the living. Along edges of all kinds, territory is being negotiated, edges and their contexts are being figured into becoming perceptible and meaningful. Mangrove claims to territory are complex negotiations with tides, storms, river flows, sediment loads, and the limits of biology and toxicity. A wall removes, or at least intends to remove, the need for ongoing relationship work as the means of maintenance of territory and replaces it with a physical barrier.[5]

In removing the need for relationship in communicating territorial power, a wall enables the, now constructed, owner of the territory to relax, to turn away from the continual effort of relationship work.[6] The power differential of territorial possession that would be worked through negotiated relationship is now built into the landscape.[7] Delineations which emerge through the fleeting, transient storyings of territoriality through intra-relationship, are transformed into definitive difference. It becomes apparently necessary, and then potentially taken for granted as part of the way things are, disappearing in the manner of Jacques Derrida's originary violence and trace, and Serres' silent lurking parasite, as discussed in Chapter 7. The power-based territorial claim may come to be taken as righteous or prior, as embedded in the physical landscape itself, as non-negotiable. The filled banks of the Cooks River become spaces that are read as "meant to be" for human determination, supporting assumptions of human exceptionalism.

By creating the possibility that territory can be asserted in a non-negotiable way, a wall offers the possibility of seeing interdependent relationships of negotiation for territory as battles that can then potentially have winners. Negotiation for territory changes from something that must be engaged in as part of everyday attentiveness, to something adversarial from which the possibility of victory is an emergent outcome. The sea wall can defeat the tsunami, the prison wall defeat crime, or the embankment defeat the flood. In transferring the power differential of territory into a physical landscape feature, the wall provides a sense of victory, of the war won, negotiation over. The wall forces one territorial claim upon the landscape and forecloses negotiation upon others, such as those made in relationship with mangroves, tides, and sea levels. The capacity for non-humans (and many humans) to negotiate territory is thus denied or seriously curtailed by walls. Lacking capacity to territorialise means being excluded from the discourses that are shaping the future.

A wall as a happy ending to a war

In his discussion of human violence towards animals, Dinesh Wadiwel picks up the theme of human sovereignty attained as a victory, further claiming that the human slaughtering and abusing of non-human animals is appropriately described as a war. He traces the development of this theme through Derrida to Martin Heidegger's claim that while all animals, all life, exert violence in their potency, humans "claim this violence as their own", as their essence.[8] It is in forgetting the act of sovereign violence that human exceptionalism and exemptionalism become narratively adopted as an assumed truth about the nature of humans. Following Michel Foucault, Wadiwel argues law enforces the sovereignty achieved through practices of slaughtering and abusing animals.[9] It is not, however, only the violence of killing that provides humans with the attitude of sovereign victory. The control gained by the structural violence of walls and other infrastructures and physical assertions of territory such as cleared land, roads, or fences also contribute. Within the storying of human exceptionalism and exemptionalism, nature and animals are arguably positioned as excluded more than they are as enemies, and against the vast numbers of domesticated animals killed by humans, we must also consider those vast numbers of non-domesticated animals who never existed because of local, regional, or global extinctions as humans seized exclusive use of territory. At some point, constraints become impossible to survive as habitat disappears or becomes inaccessible. What is done by geography, by the physical imposition of constraints, needs consideration as well as what is done by law, for the two work together. For Wadiwel, conceiving the foundation of Western human/animal relations as built upon victory in war has a far-reaching explanatory value. It explains how neither laws of protection nor those of welfare have made significant improvements to the mass cruelty towards animals because these are construed within the premise of human domination, so they contribute to "a fractal of continuing warlike domination",[10] adding detail to the structure of domination, but never fundamentally challenging it.

Building a wall is a particular territorialising action, made as a discursive move, whose goal is to shut down the conversation, conserving and establishing a perceived advantage as given. In general, territorializing can be considered a political, discursive regime, and a wall aspires to provide, for the wall owner, a "happy ending".[11] The wall as a happy ending is the sense in which a wall represents a seemingly incontestable validation of a particular relational story and of a particular strategic engagement within it. It asserts the success of the owner's strategy, as measured against the criteria of the story within which the benefits of the wall are meaningful. Donna Haraway warns us against happy endings such as these.

The colonial enterprise used walls to set its values and aspirations over the catchments and hills of the Sydney region. From the start of the settlement, the control of water was a central part of that enterprise. The Cooks River dam and embankments seized control of water diverting it towards the colonial landscape and vision and away from the local Cadigal, Bediagal, Wangal, and Kameygal people, and also from the mangrove, swamp, and forest vegetation and the animals that relied upon it. Mangroves were decimated upstream from the dam on the Cooks, and again as the estuary was reshaped by harvesting and for development. But some walls are less imposing and more qualified than others. Along the Cooks River now, councils are removing sections of the concrete channel and metal embankments and replacing them with rougher and more permeable surfaces. Mangroves are testing these for living spaces and revegetating the banks in many places. These sites offer some accommodation of the otherness of mangroves along the river, but there is little doubt about the power dynamics here. These mangroves are measured and monitored. They can, at any time, be assisted, encouraged, restrained, or removed. The morality of the colonial imagination is still highly visible in these modest invitations made to mangroves. It is easy and comfortable for a powerful sovereign host to invite a guest into their territory on their terms, but such hospitable acts nonetheless involve some move towards recognising the needs of others within a shared environment, a topic that will be discussed further in the final chapter.

Rest, directionality, clinamen, and ignorance

Rest, directionality, clinamen

Beyond the desire for security that drives so much of the shape of human patterns, of walling, of shoring up power, is a perhaps even more basic desire; the desire for rest. In offering the potential for escape from the constant negotiations, for some peace and quiet, a wall allows new things to arise within the security of the safe home, studio, deep cavern walls. Human creativity and reason can explore in new and brave ways. But what is rest? The question deserves attention. Rest neatens and makes more even, the otherwise ever-changing spaces where discernment is narrativised.[12] Rest is the pause where the violence that was brought to bear in the process of achieving that rest is able to build narratives to justify itself. Indeed, security can be thought of as the capacity and opportunity to rest without irritation. Serres identifies this in his discussion of the parasite as noise. He emphasises how the feeling that we need a break from constant interference is intense and the achievement of that rest so satisfying. "Those who control the regulation of codes and their circulation in space allow the watchers to let down their guard and sleep on the consoles of their ships, listening to music", he writes.[13] In the wall is the desire to be able to cease the constant need to negotiate, to be able to continue without relating, or at least to own the terms of the negotiation. According to Serres, maintaining a consistent directionality is itself a kind of rest. It can only be achieved by reducing or removing the diversions and distractions of parasitic interruptions. But directionality requires something like a methodology in order to persist; a methodology, an algorithm, a code, a relation, a pattern, something that might be able to repeat, that might stick sufficiently to continue, perhaps because it might be a "good enough fit". A wall forces a directionality upon the world even if it might not be such a good fit, if a wider range of desires were to be considered. The Cooks River dam and embankments forced directionality upon the water flow, upon the potential for mangroves, attempting to constrain the channel into a depth and shape more in line with the imaginations of the city merchants and developers, removing (in theory) what those powerful city fathers regarded as distractions, the parasitic interventions of mangroves, indigenous land practices, floods, and sewage from the development trajectory of the growing city.

This directionality is an emergent and unfolding relational narrative. Sometimes a pattern forms that provides the chance for some humans to become healthier, more pain free, perhaps more able to appreciate with greater intensity and delicacy their sensual experiences. One response to that emergent pattern might be to wall it, in order to maintain it and push off from it through to extend the benefit, to claim it as complete, to have and to hold. Like winching a car up hill, after each achievement, the wall needs to be built to hold the gains in place, then further gains, and another chock, another winch to go higher, and another chock. It might come to appear as though there is a goal. Such directionality provides something that can potentially continue. An experiment has begun, a methodology is in play; perhaps there is a chance to control the experiment, to establish the methodology as valid. In a wall is some desire to make a controlled experiment of life.[14]

The clinamen is one concept used to describe the fickle sensitivity of directionality. Lucretius uses the term to describe the spontaneous deviation to an atom's trajectory. It is the swerve or clinamen of the atom which enables the possibility of collision with another atom. As the uncaused origin of alterity, the original disruption of the otherwise constant and unremarkable flow of atoms along parallel paths, for Lucretius it was the source of differentiation and hence of the natural world. Gilles Deleuze and Félix Guattari,[15] Derrida,[16] and Serres[17] all pick up the concept. It fits with the importance they each give to difference or disruption as the source of patterning, though each of the writers emphasises slightly different aspects of the concept. Deleuze seeks to take it back one step; rather than being the change in direction, he finds it "the original determination of the direction of movement of the atom",[18] a directional predilection within matter at the atomic scale that enables patterning and liveliness to appear in the material world. While Deleuze's usage emphasises impetus, Serres finds a kind of primary askewness. In Serres' thinking, difference begins when something, some flow, some trajectory goes off-plan, develops a lean: "Those who limp are the discoverers; inclination is the beginning of the world".[19] For him, clinamen is the lean that causes the shift in flow, it is the "minimal angle leading to the formation of a vortex, and appears by chance in a laminar flow".[20] The lean leads to a curve, and perhaps a vortex, and hence a forming of some potentially discernible shape. Gaston Bachelard's notion of the mollusc's motto: "one must live to build one's house" comes to mind, the growing curve of the shell emerging from the ongoing living of the mollusc, who lives always with a lean.[21] Bachelard reflects "it is the formation, not the form, which is mysterious".[22] Serres' clinamen responds to that mystery; "The parasite is an operator; it is a generalized clinamen".[23] Derrida's usage emphasises the chancy, unfathomable quality of this originating concept. For Derrida, the clinamen assists in his endless interrogation of the processes of semiotic material writing; an interrogation which is interwoven throughout his work with other threads tracing the same repeating pattern of (seemingly) impossible and unpredictable generation from a situation of fundamental aporia. Within the aporia, the clinamen is the element of chance, the "who knows what or how" that suggests an inclination towards one thought rather than another, one word rather than another, and so on. "My clinamen, my luck or my chances, incline me to think the clinamen beginning with the divisibility of the mark".[24] The mark here refers to Derrida's own written marks, and also to all other "indivisible elements, atoms" of linguistic systems,[25] including those beyond the human. In all cases, he suggests, there is an aporia involving indivisibility and yet also deviation, or insignificance (each instance of a mark is only itself – it has no parts, is not connected within a meaning system) and remarkability (it must be repeatable across instances). This iterability "imprints a power of diversion on its very movement. It is, in the destination, a principle of indetermination, chance, randomness".[26] If the wall is a desire to stop time at a propitious moment, where one is powerful, before another event, another divergence occurs to reorganise the cosmos, it can be understood as an attempt to control the clinamen, to deny chances, to stop the diversion from a particular unremarkable trajectory.

Ignorance

In providing an opportunity for rest, walls facilitate ignorance as an indicator of success. For those whose security is produced by the infrastructure, there is no longer any need to see or smell, let alone become familiar with the mangroves, the homeless, the slaughterhouse, the loss of life that has occurred to provide this security. The focus of attention can be dramatically reduced. The more successful the wall is at closing off space and stopping time's multiplicities, the less those who are advantaged by it need to notice. Irritations and challenges to the closure and coherence of the narrow, human-oriented frame of reference that has been crafted at such expense can be ignored. If we are very secure, with a secure room in a secure house in a secure neighbourhood in a secure city in a secure state, we can sleep undisturbed and spend our days attending a small screen right in front of us. Inattentiveness can therefore function as a status indicator, demonstrating attunement between the inattentive person's goals and those that are the status signifiers of the system. Serres notes this alignment of status with ignorance when he writes, "whoever belongs to the system perceives noises less and represses them more, the more he [sic] is a functioning part of the system".[27] When status aligns with not paying attention, there is normative social pressure on people to disengage with their surroundings, particularly with aspects evaluated as excluded.

As attentiveness within relations is fundamental to thriving within a complex ecology (as discussed in Chapter 6), some thinkers have noted the importance of being aware of factors, such as this normative pressure towards ignorance, that push in the opposite direction and have suggested material semiotic approaches to countering them. Isabelle Stengers draws attention to this tendency towards inattention within the context of the enclosures of capitalism, arguing that "free enterprise has conquered the right not to pay attention".[28] She urges her readers to refuse the walls of capitalist security, and to seek methods to help reposition ourselves as vulnerable in order to create the opportunities we need to pay attention better within broader ecologies of practice. To do this, she suggests we devise and employ narrative tools to support the maintenance of greater vulnerability within our social practices and their politics. She proposes several such narrative tools, including paying attention to the intrusion of an implacable Gaia, seers, tarot, diversity requirements, and variations upon the role of the idiot.[29] Others point to the role of resistance by those who are marginalised. Wendy Brown explains that a wall: "facilitates denial of the dependency of the privileged on the exploited and of the agency of the dominant in producing the resistance of the oppressed".[30] Walls divert resources from and repress the opportunities of those who might challenge and resist the would-be self-sufficient invulnerable subject. Refuges for the repressed and excluded, offering opportunities for survival and resistance, become critical.

Chapter 9 will continue this discussion to consider the role of the Cooks River mangroves within a narrative of post-colonial resistance.

The architect Bernard Cache offers a contrasting view that presents the reasoning and assumptions underlying the position Serres, Stengers, and Brown are countering. He assumes, with a humanist, rationalist optimism, that imposing particular privileged human intention upon a chaotic world by building walls, windows and roof is indeed a route towards a desired better future. For Cache, it is architecture, "the art of introducing intervals in a territory in order to construct frames of probability", that must create the spaces required for human possibilities.[31] The architect, he suggests, must minimise the ambiguity of the intervals, which are rooms and gaps, to maximise the architect's desired effects; "one can increase the probability of an expected effect by eliminating all disturbances in the interval".[32] Cache makes no pause to consider the source or implications of the authority with which the privileged human assumes the right to design territory, including "eliminating all disturbances" as they see fit. In contrast with Serres, Cache finds the simplifying, flattening, and emptying of space to be a requirement for human creativity: "It is the flatness of the stage that makes choreography probable, just as it is the flatness of the stadium that increases the probability of athletics".[33] Here is the wall working in partnership with the creative aspirations of elite privilege, even despite an awareness of the liveliness and potentialities of space and time. The generative spaces of gaps are places that can be colonised, that are continually colonised. Mere awareness of their potential does nothing to reveal the politics and power relationships at play. Flattened and bordered spaces certainly facilitate dance and sport, and that is not to be dismissed. The issue here is that in believing that human fulfilment requires the enclosing and emptying of space, and in assuming the sovereignty of a privileged elite to impose their version of that belief as one limiting vision, the future possibilities of innumerable others, human and non-human, are entirely overlooked.

Along the Cooks River, I seldom see a jogger, cyclist, or player of ballgames who notices the mangroves. The historian J. D. Dunstan provides this summary of the attitudes of the planners and engineers in the Sydney region through the early to mid-20th Century:

> Municipal and shire engineers and planners, competent in their own fields, often viewed wetlands and waterways as cheap land to be filled and reclaimed, or as a source of commercial sand. The backwaters and mangrove swamps offered the ideal location for the disposal of garbage with little thought given to the tragic effects on the natural environment. In the Municipality of Hurstville, bordering the Georges River, some 80 per cent of the present open space was obtained from reclamation of mangrove swamps, primarily using such areas for garbage disposal. When in 1969 objections were raised against further reclamation

> to extend a golf course, the Shire Engineer was reported in the press as saying: "Mangroves are a noxious weed and a cancerous growth". There was no public objection to this statement.[34]

People threw things that they wanted to disappear into these mangroves because they were regarded as an "away", a place beyond the urban, human domain. They became a site for the illegal or the toxic. The Cooks River carried the organic and inorganic wastes from industry and residential uses for many decades, becoming a maligned and noxious waterway. Tyrrell notes that "by 1975 there were 2100 businesses capable of doing damage, including smelters, paint and chemical manufactures, dye works, motor vehicle workshops and electroplaters – all connected by drains to the river and discharging into it".[35] When mangroves returned, finding sites to thrive despite the embankments and toxins, they performed what Stacey Alaimo describes as "exposure as insurgent vulnerability",[36] occupying this difficult, poisoned in-between in ways that strike hope for more-than-human Cooks River futures.

Embankment

Cleaning, emptying, simplifying

Even before walls, there is the sweeping of dirt, the clearing of ground, the cleaning up of filth. Across the globe, women are sweeping; sweeping around campfires, sweeping courtyards, sweeping pathways, sweeping floors of all kinds. For Serres, this desire to clean is fundamental to the establishment of cultures:

> the deparasiting of a finite location. This is the oldest trick in the book, the most common in history, the most reproduced in our institutions, cultures, and sciences. Cut, center, purify [...]. Maybe it is the fundamental model or structure of producers' cultures.[37]

Serres comes to think that this cleaning is ultimately there behind all the labours of human societies, necessary for human social structure. The desire to clean can be experienced as a powerful inclination, and one supported in many contexts by deeply culturally embedded norms. Excluded when we clean are the other beings whom we consider uninvited; dust and dust mites, mud, insects, the putrid which teems with bacteria and fungus.

When the uninvited other is something as uncompromising as cold, wind, rain, tide, or flood, there may seem to be little recourse to negotiation, and sweeping is not enough. In response to these irritants, it seems, a wall must be built. Once there is a house wall, and an inside, it can be radically cleaned, the aim can be to empty it out. In a clean and empty space, time slows and chatter quietens. Serres writes: "The house functions as a space of transformation where forces are calmed, like a high energy filter, or converter".[38] This quietened, slowed space suggests there might be states of optimal well-being where a human might be pure and free of pollution, invasion, or disruption.[39] It suggests that it might be possible for narratives such as a human life, a family narrative, a national history narrative, or histories of human-only achievement to be worked through protected from negative external influences. It suggests that there might be sufficient simplicity, closure, and control for rationality to operate in decision-making and economies. All of these suggestions strengthen the supposition of the separability of human goals from the goals of all others, and of human exceptionalism.

Never adequately achieved, however, the work of emptying out is endless and demanding. We, the wall builders, the possessors, the hosts, can never quite manage to have the last word – there will be interruption, at the very least, noise, wind, cold, our own gut bacteria demanding to be fed. Serres paints a tragic picture of the endless labour of cleaning out that ever seeks the emptiness required to quiet the parasite; "We will finally be at home when everything is clean".[40] In terms of the prescient self, perhaps the hope is that all that cleaning might make the self empty enough to find out what it "truly" is, in a state entirely free from parasites. Serres claims that desire for a state of purity – in its ultimate form a search for pure reason – becomes an addiction, with escalating costs[41] and tragic consequences; "the dead pile up to the very end. The tragic is the flood and the blank plain".[42] Brown explores this in her discussion of the roles of walls in maintaining "fantasies of purity, innocence and goodness",[43] and in his history of the Cooks River, Tyrrell recounts that removing mudflats and creating clear water and solid land was an ongoing and expensive struggle along the Cooks River for generations, the elusive dream reflected in the title of his work; *River Dreams*.[44] As colonial infrastructure, the Cooks River dam wall and embankments were part of a purification narrative that repurposed the swamp as a drainage system for industrial and urban waste and as riverside recreation areas and negated the mangroves and the Indigenous humans. The mangrove's return was visible and controversial. It upset the purity narratives and began an alternative more-than-human riverbank conversation, which will be discussed in Chapter 9.

"Washing is a social act; purifying one's space is an act of welcoming, or a religious, amorous, collective, or hostal act", Serres explains.[45] It is also hostile; washing involves acts of excluding and killing that enable the acts of welcoming-in. Few social criteria are as deeply prescriptively normative as those of cleanliness. Cleaning confronts us with dilemmas pitting social norms against personal ethics. I have often experienced these in the daily situations of sharing a house, neighbourhood, and society with others; the small stream of ants across the kitchen bench expands into the pantry, the unmown lawn is abundant with butterflies and bees, the reluctance to use detergents results in clothing that appears less clean than that of peers.

Killing the ants or mowing the lawn confront as opposed to the way I wish to live with others, yet at some point the situation becomes a source of social shame. In these situations, I reluctantly kill or pollute in order to present as a worthy human within a social hygiene narrative I do not respect but have often been unable to escape. Holding out against social hygiene norms takes a toll on social status and the associated opportunities it enables, and demands committed political effort to continually present publicly what is normatively judged to be poor behaviour. One of the promises of increasing affluence is relief from these confrontations as achieving stronger walls might mean a house less penetrable to ants or paving and garden beds in place of the grass. Walls seem to promise an escape from the messy moral dilemmas of sharing spaces, but all my daily violences are transferred into their material forms and while I may be relieved, the ants and bees are defeated.

The claim made, the territorial power asserted, the wall built, inside and outside become possible to conceive. Building a wall enables the simplification of both inside and outside. Both can become less attended to, but in particular and very different ways. From the perspective of the border-makers and keepers, the sovereign hosts at the gates, inside becomes simple and vague because it is protected. Only its boundaries require attention, the details, the border-keepers may assume, can become the obsessions of those who, because they are protected inside, have no other more worldly cares – the women in a patriarchal society, for example – who can become ignored inside.[46] Outside can also become less attended to. As Bachelard describes, there can now come into being a very big vague thing called "outside". He writes: "Outside the occupied house the outside is a simplified cosmos. It is a non-house".[47] When I am inside, the entire world beyond the walls of my domain can potentially become one simple space with barely any detailed features. Grosz enlarges on this theme, exploring the creation of the concept of landscape as the framed outside:

> While its most direct and perceptible function is to separate or divide, the wall equally functions to select and bring in. In this case, the frame can be converted into the window, which selectively envisions its natural exterior, now a "landscape," no longer beyond its partition but within the enframed space of the room.[48]

The metaphor of inside and outside has profoundly influenced concepts in many languages: "metaphysics has a geography of inside/outside", Bachelard writes.[49] The house walls, like the walls of human exceptionalism and exemptionalism, establish dominions which can be filled by the owner's narratives, such that no others can find space to establish themselves.[50]

The mangroves, the mosquitoes and the mud were cleaned away along the Cooks River. The wide river margins contracted to nothing as concrete and metal edges separated water from land. Margins, however, provide essential services for a growing city, they are places of refuge, places to throw the things that have no allocated places, sources of imagination and, in the case of the Cooks River mangroves, of challenge to the colonial enterprise of water and land control. In returning, the mangroves offered a challenge to the dreams of colonial values.

Decision

Can we make decisions when we are ignorant and cut-off inside walls? Or can we make decisions only when we have the rest we need from constant interruption? While the ignorance and inattention that walls enable limit the evidence base for decision-making, the security and constraints they provide support the possibility of extended focus and rigorous analysis. Serres is firmly on the side of stepping outside the walls. He is critical of decision-making occurring in an inadequately dangerous context, decision-making that has been protected by walls and where situations are encountered only through the frames of windows (or human-centred value systems).[51] Serres is a philosopher against security and for alertness to precarity as a force for ethical attentiveness. The world outside protective walls is "organized and defined by its constant diligence".[52] Here nerves respond to stimulus without protection and living requires vigilance. In the unwalled world we are constantly called to account by death. Serres holds that the diligence demanded of us in a space where death is a constant presence is required for responsible decision-making.[53]

For Derrida, the act of making a decision occurs inside an impossible space that always involves a leap. Rational decision-making requires knowledge, but that can never be sufficient for certainty, so "the moment and structure of the 'il faut' of the 'it is necessary' […] must remain heterogenous to knowledge". An absolute interruption must separate them.[54] This makes decision-making "both an act of an order, a fiat, a decree, and as a schism, a caesura, a separation, a dissection".[55] He emphasises that the schism is internal, a necessary structural element within the act of deciding, indeed of meaning, which relies upon iterative and interdependent decisions; "a cleavage and torment interior to meaning in general, interior to logos in general, a division within the very act of sentire".[56] This impossible but necessary madness thrusts decision makers into both responsibility and incalculable risk, and, Derrida claims, is the driving force of history.[57]

Serres' decision maker is more or less responsible, indeed more or less "real" depending upon their degree of vulnerability and interconnectedness. He does not suggest that a good decision will emerge from being entirely alert, attentive, and aware of ones's vulnerability, or even that it cannot emerge from a more comfortable situation, but that this is the best position from which an apt action is likely to emerge. The further from this state, the more protected and inside, the less likelihood there is of due attentiveness. Thus, for Serres, the context of the decision-making is always important.

For Derrida's decision maker the distinction between the vulnerability and security of the context is less relevant. The protective walls of a house or a nation make no difference because the space of the decision is more abstract, with a generalisable structure unrelated to specific context. Any decision finds one facing the impossible, having to leap into the unknown "a decision has to go through some impossibility in order for it to be a decision".[58] The wall is now inside the decision process, an impasse that must be breached, yet cannot be breached without abandonment to the other. Erinn Gilson explains it like this:

> to experience the aporia is not to attempt to solve or move beyond the impasse it presents but rather to remain within the impasse – immobile, enduring the lack of knowledge with which one is presented. [...] One attempts to move and think within impossibility, in an experience of endurance that is not one of simple passivity but is in fact the condition of decision and responsibility. The unending undecidable character of aporia make responsible decision possible, makes thought possible.[59]

In the material semiotic context, a responsive move, as the mosquito turns towards one wallaby or another, as the oyster genotype becomes more precisely attuned towards a phenotype that can sort detritus in a way that extracts sufficient nourishment, as the mangrove tree refines its capacity to exclude salt, is a decision. Decisions among us as our relations unfold are therefore abundant, take highly diverse material semiotic forms, and operate at varied durations and scales. Derrida suggests decision-making is inescapable within every interaction, while Serres suggests its challenges are most responsibly engaged with by attending fully to each interaction's demands. These positions both accord well with those views of responsibility associated with relational ontologies discussed in Chapter 4. Decision-making processes are differentiation processes, the same fractal foldings generating criteria of reference are at play. How mangroves make walls, and approaches to collaborative wall-building between mangroves and humans along the Cooks River will be discussed in Chapter 9.

Walls make a particular difference appear essential, instead of disputed in negotiation between. This occurs as walls replace the need for attentive and responsive relationship work in figuring difference at its many interfaces. This makes it possible to read territorialising as something that can have an end, readable as a competition where one strategy can prevail, and indeed be claimed to have done so. A wall can appear to remove violence and pain from their places at the interface of affect, as they occur within the negotiations of difference, while the considerable violence of the imposition and maintenance of the wall is perpetuated within its architecture.

Notes

1 Michel Serres, *The Parasite*, trans. Lawrence Schehr (Minneapolis: University of Minnesota Press, 2013), Meals of the Lord in Paradise, paragraph 21, Kindle.

2 Tema Milstein argues ecocultural scholarship must examine the social structures sustaining the norms that "patrol the boundaries of dominant anthropocentric identity" and deter ecocultural identities in "Ecocultural Identity Boundary Patrol and Transgression," in *Routledge Handbook of Ecocultural Identity*, ed. Tema Milstein and José Castro-Sotomayor (London: Routledge, 2020), 27.

3 Stuart Elden, *The Birth of Territory* (Chicago and London: The University of Chicago Press, 2013), 17, Kindle.

4 In reflecting upon Shirley Strum's observations of the society-building and maintaining work of baboons, Bruno Latour comments:

> If sociologists had the privilege to watch more carefully baboons repairing their constantly decaying "social structure", they would have witnessed what incredible cost has been paid when the job is to maintain, for instance, social dominance with no thing at all, just social skills,

in *Reassembling the Social: An Introduction to Actor Network Theory* (New York: Oxford University Press, 2012), 70.

5 Viveiros de Castro recalls

> the final paragraph of *The Elementary Structures of Kinship*, where Lévi-Strauss observes that in both the myths of the Golden Age and beyond, "mankind has always dreamed of seizing and fixing that fleeting moment when it was permissible to believe that the law of exchange could be evaded, that one could gain without losing, enjoy without sharing," and that total happiness, "eternally denied to social man" would consist in "keeping to oneself",

in *Cannibal Metaphysics*, trans. Peter Skafish (Minneapolis: Univocal, 2014), chapter 7, paragraph 12, Kindle.

6 In the terminology of information theory, the relationship between walls and the power differentials of territory might be translated in terms of that between structure and information, or syntax and semantics; the more work done by the structure, the more efficient can be the use of the information. The comparison between poetry and prose is an example – poetry employs structure more intensely than prose and achieves greater efficiency in its use of the language.

7 John Latour writes:

> It's the power exerted through entities that don't sleep and associations that don't break down that allow power to last longer and expand further – and, to achieve such a feat, many more materials than social compacts have to be devised,

in *Reassembling*, 70. A similar comment is made in "Actor Network Theory and Material Semiotics," in *The New Blackwell Companion to Social Theory*, ed. B. S. Turner (Oxford: Blackwell, 2009), 148.

8 Dinesh Wadiwel, "The Will for Self-Preservation: Locke and Derrida on Domination, Property and Animals," *SubStance* 43, no. 2 (2014): 158.

9 Dinesh Wadiwel, "The War against Animals," *Griffith Law Review* 18, no. 2 (2014): 287, https://dx.doi.org/10.1080/10383441.2009.10854642
10 Wadiwel, "War", 290.
11 This is a reference to the quotation from Donna Haraway included in the "Attunement" section of this book in Chapter 5: "It's not a 'happy ending' we need, but a non-ending", "The Promises of Monsters: A Regenerative Politics for Inappropriate/D Others," in *Cultural Studies*, eds. Lawrence Grossberg, Cary Nelson, and Paula A. Treichler (New York: Routledge, 1992), 327.
12 Delays, such as those provided by rest, are essential according to Bergson's process ontology where becoming emerges from a series of steps of activities and delays. The delays allow connections to be made, for the structures of discernment to become stabilised. For a discussion of delay in Bergson and Deleuze, see Claire Colebrook, *Deleuze: A Guide for the Perplexed* (New York: Continuum, 2006).
13 Michel Serres, *The Five Senses: A Philosophy of Mingled Bodies*. Translated Margaret Sankey and Peter Cowley (New York and London: Bloomsbury, 2016), 'Hermes and the Peacock', paragraph 64, Kindle.
14 I note that a laboratory is one of the most walled of spaces. Some laboratory walls keep out almost everything, and almost all ways of thinking and relating and experiencing, to hold all this away, so the experiment can be run without interruption.
15 Gilles Deleuze and Felix Guattari, *A Thousand Plateaus: Capitalism and Schizophrenia*, trans. Brian Massumi (Minneapolis and London: University of Minnesota Press, 1987).
16 Jacques Derrida, *Psyche: Inventions of the Other*, eds. Peggy Kamuf and Elizabeth Rottenburg, vol. 1 (Stanford: Stanford University Press, 2007).
17 Serres, *Parasite*.
18 Gilles Deleuze, *Logic of Sense*, trans. Mark Lester (London: The Athlone Press, 1990), 269.
19 Serres, *Parasite*, Athlete's Meals Difference and the Construction of the Real, paragraph 9.
20 Serres, *La Naissance de la Physique* at 14, cited in Deleuze and Guattari, *Plateaus*, note 17.
21 Gaston Bachelard, *The Poetics of Space* (Boston: Beacon, 1969), 106.
22 Bachelard, *Poetics*.
23 Serres, *Parasite*, Cows Eat Cows, Theory of the Line, paragraph 27.
24 Derrida, *Psyche*, 360.
25 Ibid., 358.
26 Ibid., 360.
27 Serres, *Parasite*, The Means, the Milieu, paragraph 8.
28 Isabelle Stengers, *In Catastrophic Times: Resisting the Coming Barbarism*, (Ann Arbour: Open Humanities Press and Meson Press, 2015), 151. http://dx.medra.org/10.14619/016
29 These narrative strategies for use within relations are developed through all of Stengers' works cited in this book.
30 Wendy Brown, *Walled States, Waning Sovereignty* (New York: Zone Books, 2010), Fantasies of Purity, Innocence and Goodness, paragraph 3, Kindle.
31 Bernard Cache, *Earth Moves: The Furnishing of Territories*, Cambridge and London: Massachusetts Institute of Technology, 1995, 22.
32 Cache, *Earth Moves*, 24.

33 Ibid.
34 D. J. Dunstan, "Some Early Environmental Problems and Guidelines in New South Wales Estuaries", *Wetlands Australia Journal* 9, no. 1 (2009): 1.
35 Ian Tyrrell, *River Dreams: The People and Landscape of the Cooks River* (Sydney: University of New South Wales Press, 2018), chapter 4, Noxious River, Kindle.
36 Stacey Alaimo, *Exposed: Environmental Politics and Pleasures in Posthuman Times* (Minneapolis and London: University of Minnesota Press, 2016), Introduction, paragraph 6, Kindle.
37 Serres, *Parasite*, Energy, Information, paragraph 4.
38 Serres, *Five Senses*, Cells, paragraph 30.
39 As critiqued by Alexis Shotwell, *Against Purity: Living Ethically in Compromised Times* (Minneapolis and London: University of Minnesota Press, 2016).
40 Serres, *Parasite*, Insects' Meals, paragraph 4.
41 He writes, for example: "Pure reason is inflation. Inaccessible purity raises its prices", at ibid., Insects' Meals, final paragraph.
42 Ibid., Impostor's Meals: Analyse, Paralyze, Catalyze, paragraph 27.
43 Brown, *Walled States, Waning Sovereignty*, Fantasies of Purity, Innocence and Goodness, paragraph 3.
44 Tyrrell, *River*.
45 Serres, *Parasite*, Salad Meals: Stercoral Origins of Property Rights, paragraph 20.
46 Wendy Brown, in her extensive work on the walls of nation states notes this gendered quality; "There is a notably gendered inflection to walling as a defence against anxieties about need, vulnerability, and penetrability and to the desire for sovereign containment and protection against such vulnerability", in *Walled States, Waning Sovereignty*, The Psychoanalysis of Defense, paragraph 17.
47 Bachelard, *Poetics*, 40.
48 Elizabeth Grosz, *Chaos, Territory, Art: Deleuze and the Framing of the Earth*. (New York: Columbia University Press, 2008), 14.
49 Bachelard, *Poetics*, 212.
50 Per Aage Brandt postulates that walls shape collective living and hence culture:

> What is "in here" belongs to "us", and what is "out there" does not (yet). The social pronoun "we" is parietal: it refers to walls. The interiorizing wall and the deictic window shaped our communitary, collective living and thinking. Structurally, they shaped a cognitive semiosis that generated culture.

In Per Aage Brandt, "The Principle of the Wall – or How Space (May Have) Shaped Semiosis. A Note", (2012), http://dx.doi.org/10.2139/ssrn.2304745
51 Michel Serres, *The Natural Contract* (Anne Arbor: The University of Michigan Press, 1995), 29.
52 Ibid., 111.
53 Ibid., 112.
54 Jacques Derrida, *Rogues: Two Essays on Reason* (Stanford: Stanford University Press, 2005), 145.
55 Jacques Derrida, *Writing and Difference*, trans. Alan Bass (London and New York: Routledge, 2001), 45.
56 Ibid. "Sentire" might be roughly translated as sensing or feeling.
57 Ibid., 50.

58 Mark Dooley and Richard Kearney, "Hospitality, Justice and Responsibility: A Dialogue with Jacques Derrida," in *Questioning Ethics: Contemporary Debates in Continental Philosophy*, ed. Mark Dooley and Richard Kearney (London: Taylor and Francis, 2002), 66.
59 Erinn Cunliffe Gilson, "Becomings yet to Come: Thought as Movement in Derrida and Deleuze," *Philosophy Today* 51, no. 4 (2007): 386, www.pdcnet.org/philtoday/content/philtoday_2007_0051_0004_0383_0392

References

Alaimo, Stacey. *Exposed: Environmental Politics and Pleasures in Posthuman Times*. Minneapolis and London: University of Minnesota Press, 2016, Kindle.

Bachelard, Gaston. *The Poetics of Space*. Boston: Beacon, 1969.

Brandt, Per Aage. "The Principle of the Wall – or How Space (May Have) Shaped Semiosis. A Note." (2012). http://dx.doi.org/10.2139/ssrn.2304745

Brown, Wendy. *Walled States, Waning Sovereignty*. New York: Zone Books, 2010, Kindle.

Cache, Bernard. *Earth Moves: The Furnishing of Territories*. Cambridge and London: Massachusetts Institute of Technology, 1995.

Colebrook, Claire. *Deleuze: A Guide for the Perplexed*. New York: Continuum, 2006.

Deleuze, Gilles. *Logic of Sense*. Translated by Mark Lester. Edited by Constantin Boundas. London: The Athlone Press, 1990. Originally published as *Logique du Sens*, Les Éditions de Minuit, 1969.

Deleuze, Gilles and Felix Guattari. *A Thousand Plateaus: Capitalism and Schizophrenia*. Translated by Brian Massumi. Minneapolis and London: University of Minnesota Press, 1987. Originally published as *Mille Plateaux: Capitalism et Schizophréne 2*, Les Éditions de Minuit, 1980.

Derrida, Jacques. *Psyche: Inventions of the Other*. Edited by Peggy Kamuf and Elizabeth Rottenburg, vol. 1. Stanford: Stanford University Press, 2007. Originally published as *Psyché: Inventions de L'autre, Tomes I et II*, Éditions Galilée, 1987.

———. *Rogues: Two Essays on Reason*. Translated by Pascal-Anne Brault and Michael Nass. Stanford: Stanford University Press, 2005. Originally published as *Voyous: Deux Essais sur la Raison*, Éditions Galilée, 2003.

———. *Writing and Difference*. Translated by Alan Bass. London and New York: Routledge, 2001. Originally published as *L'écriture et de la Différence*, Éditions du Seuil, 1967.

Dooley, Mark and Richard Kearney. "Hospitality, Justice and Responsibility: A Dialogue with Jacques Derrida." In *Questioning Ethics: Contemporary Debates in Continental Philosophy*, edited by Mark Dooley and Richard Kearney, 65–83. London: Taylor and Francis, 2002.

Dunstan, D. J. "Some Early Environmental Problems and Guidelines in New South Wales Estuaries." *Wetlands Australia Journal* 9, no. 1 (2009): 1–6.

Elden, Stuart. *The Birth of Territory*. Chicago and London: The University of Chicago Press, 2013, Kindle.

Gilson, Erinn Cunliffe. "Becomings yet to Come: Thought as Movement in Derrida and Deleuze." *Philosophy Today* 51, no. 4 (2007): 383–92. www.pdcnet.org/philtoday/content/philtoday_2007_0051_0004_0383_0392

Grosz, Elizabeth. *Chaos, Territory, Art: Deleuze and the Framing of the Earth*. New York: Columbia University Press, 2008, Kindle.

Haraway, Donna. "The Promises of Monsters: A Regenerative Politics for Inappropriate/D Others." In *Cultural Studies*, edited by Lawrence Grossberg, Cary Nelson, and Paula A. Treichler, 295–337. New York: Routledge, 1992.

Latour, Bruno. *Reassembling the Social: An Introduction to Actor Network Theory*. New York: Oxford University Press, 2012.

Law, John. "Actor Network Theory and Material Semiotics." In *The New Blackwell Companion to Social Theory*, edited by B. S. Turner, 141–158. Oxford: Blackwell, 2009.

Milstein, Tema. "Ecocultural Identity Boundary Patrol and Transgression." In *Routledge Handbook of Ecocultural Identity*, edited by Tema Milstein and José Castro-Sotomayor, 26–52. London: Routledge, 2020.

Serres, Michel. *The Five Senses: A Philosophy of Mingled Bodies*. Translated by Margaret Sankey and Peter Cowley. New York and London: Bloomsbury, 2016, Kindle. Originally published as *Les Cinq Sens: Philosophie des Corps Mêlées I*, Grasset et Fasquelle, 1985.

———. *The Natural Contract*. Translated by Elizabeth MacArthur and William Paulson. Ann Arbor: The University of Michigan Press, 1995. Originally published as *Le Contrat Naturel*, Editions Francois Bourin, 1992.

———. *The Parasite*. Translated by Lawrence Schehr. Minneapolis: University of Minnesota Press, 2013, Kindle. Originally published as *Le Parasite*, Grasset et Fasquelle, 1980.

Shotwell, Alexis. *Against Purity: Living Ethically in Compromised Times*. Minneapolis and London: University of Minnesota Press, 2016, Kindle.

Stengers, Isabelle. *In Catastrophic Times: Resisting the Coming Barbarism*. Translated by Andrew Goffey. Ann Arbour: Open Humanities Press and Meson Press, 2015. Originally published as *Au Temps des Catastrophes: Résister à la Barbarie qui Vent*, Editions La Découverte, Paris, 2009. http://dx.medra.org/10.14619/016

Viveiros de Castro, Eduardo. *Cannibal Metaphysics*. Translated by Peter Skafish. Minneapolis: Univocal, 2014, Kindle. Originally published as *Métaphysiques Cannibales*, Presses Universitaires de France, 2009.

Wadiwel, Dinesh. "The War against Animals." *Griffith Law Review* 18, no. 2 (2014): 283–97. https://dx.doi.org/10.1080/10383441.2009.10854642

———. "The Will for Self-Preservation: Locke and Derrida on Domination, Property and Animals." *SubStance* 43, no. 2 (2014): 148–61. https://doi.org/10.1353/sub.2014.0029

9 Mangrove walls, mangrove hosts

More-than-human hospitalities

Michel Serres famously reminds us to notice swamps as actors, "aren't we forgetting the world of things themselves, the sand, the water, the mud, the reeds of the marsh?"[1] This chapter pays attention to the way the Cooks River mangroves work the riverbanks and function, in many respects, as walls. Their ways of walling are remarkably different from those of the human embankment builders, occurring through processes accommodating, accumulating, and diversifying rather than processes of excluding and simplifying. I explore the possibility that mangrove ways of walling can suggest alternatives to the violence and imposition of the walls of human exceptionalism, both materially, and for ways of thinking and relating more broadly.

Mangroves as walls

Porous mangrove walls

On 26th December 2004, an earthquake induced tsunami struck Asia's Indian Ocean coasts killing more than 200,000 people, predominantly in Indonesia's Aceh province and Eastern Sri Lanka. Research in the aftermath was quick to identify profound differences in the extent of destruction between coastal areas where mangroves were still present, and those where the mangroves had been removed for urban development and shrimp and fish farms. Several studies reported the protective role that mangroves played where they were present and in a healthy condition.[2] The general proposition that mangroves protect coastal areas from damage during violent storms has now been widely accepted,[3] with estimates that mangrove ecosystems reduce wave energy by 50%.[4] Mangroves present a massive but permeable barrier to storm surges or tsunamis, absorbing much of the energy of the incoming water into their structures.[5] Additionally, the mud islands and banks of accumulating sediment and biomass from roots, leaves, and algal mats can form living walls protecting the coastal areas against sea level rise.[6] Mangroves can potentially build up mud faster than the predicted rate of sea level rise. Because of these wall-like effects, replanting and conservation of mangroves, particularly across the tsunami-affected countries

DOI: 10.4324/9781003286493-13

have increased greatly and there is a growing recognition of the importance of mangroves for coastal protection.[7]

It is not only in protecting coastal integrity against the destructive force of the oceans that mangroves act as walls. Mangroves also intercept pollutants, suspended matter, and excess nutrients running into waterways from the land, slowing these from entering the rivers and oceans and accumulating them within their mud.[8] In this way, in many areas mangroves are mitigating estuarine eutrophication caused by effluent from fish and shrimp ponds.[9] Community volunteer groups and groups of offenders on community corrections orders remove over a hundred tonnes of rubbish from the Cooks and Georges each year,[10] much of it accumulating in the mangroves as storm water carries it through the drains into the mangrove-lined inlets. Mangroves also create protected niches for the young of many fish species and for shellfish and crustaceans.[11] In areas of Aceh province where mangroves were regenerated around surviving fish and shrimp farms, the productivity of these increased, so now models of farming that include mangroves are becoming popular.[12]

In protecting coasts, providing "insides" for fish spawn, and in their role as storage, mangroves function as walls. But their ways of walling are very different from the walls of sovereign power. They are highly porous, protecting against the storm's violence not by stopping it suddenly with massive force but by dispersing its energy and damage through their bodies. They provide protection not by enclosing but by offering niches of opportunity and sanctuary within themselves. They build up sediment as sea levels rise not by destructive and expensive movements of resources, but by working with the flowing tides and suspended particles to slow them and settle them into the mangroves own gathering body. They become rich carbon stores because there is such a gathering together of life within them, deep into the mud where so much falls and settles. Instead of excluding and stopping, or containing and keeping, the mangal provides the functions of a wall by absorbing and accommodating. It offers protection through its porosity, offers defence through its complexity of structure.

Mangrove bank building is performed by diverse biological and geomorphic processes. Flocculation is the process by which suspended mud particles bind together into heavier aggregates that can settle out from a waterborne state. It involves the intertwining of hydrological, biological, and geological factors. Tiny grains of suspended marine sediment floating in the current become home for many bacteria. Each grain of marine sediment hosts between 10^4–10^5 microbial cells[13] so these grains are a complex intermingling of mineral (or micro plastic) and biological. The bacteria secrete sticky mucus, so as their movements slow in the calmer estuarine environment, the particles can bind together into larger, heavier flocs that can settle out of suspension. Flocculation is greatly aided by oysters and other filter-feeders who suck large quantities of marine particles through their bodies. The indigestible grains of sediment are caught in a ribbon of slime and moved out of the body as pseudofaeces.[14] These larger, heavier, slime-coated grains flocculate together even more quickly, clearing the water as it flows through the oyster reefs towards the adjacent river mangroves. Mangrove roots grow down into the nourishing mud bank and stabilise it. Algal threads, bacterial films, and fungal mycorrhiza also help hold the mud together, like a biocement for the mangrove walls.[15] The height and volume of the muddy mangrove walls are increased by large quantities of voids within it built by many burrowers such as crabs, crayfish, shrimps, and worms. Mangrove tree roots growing through the mud also contribute to the volume and to the capacity of mangrove mud banks to rise and to become further inhabited.

Semiotic interpretive processes enfold to allow a mangal mud bank to be meaningful as a wall. The mangal is interpreted as a restrictive or protective barrier within many different behavioural practices and physical responses. In Charles Sanders Peirce's pragmatist semiotics *that* border, the wall, is "mutual expectations of behaviour, which [...] tend to be carried out by certain real and hard physical entities".[16] Peircean semiotics finds meaning written by the actions of the material world, living and non-living – flows of water shaping the river banks, for example,[17] and follows those "hard" physical actions as they become habitualised into "soft" social or linguistic forms.[18] The soft folds back into the hard as well, as attitudes (for example) affect behaviours. A mangrove wall is lived as mutual expectations of the behaviours of all those (living and non-living) who are stopped by its body or who seek shelter behind it.

Membrane

Some studies are revealing that permeable barriers such as large expanses of mangroves minimise storm damage more effectively than massive solid concrete barriers.[19] While the mangroves allow more water to flow through them and onto the land beyond than the solid massive wall, the water will be calmed and not destructive. A massive sea wall, if it is strong and high enough, can certainly stop the waves from overflowing the shore, but many sea walls fail, removing all protection. Further, coasts erode away from around and under sea walls, eventually undermining them.[20] As a result, porous sea walls that operate more like mangroves or marshes, softening and absorbing the incoming tides and storm surges are now gaining interest.[21]

A selectively permeable wall that supports semiotic processes of sorting between two areas is a membrane. In her semiotic material discussions of membranes, Astrida Neimanis contrasts the membrane with that which separates and severs. The membrane, she argues, enables differentiation, but in a way that is concerned about the nuances of the processes of the between space, incorporating the communicative negotiations of territorialising within the dividing barrier. Where difference is brutally forced by a wall, it is enabled by a membrane as an ongoing process of communication between.

At the edges where watery body meets watery body there are, she writes, "shared cyclings".[22] The Cooks River mangroves are membranous and gestational. Mud and flow bring together things whose interactions are generative as well as destructive, as illustrated through the key examples of this book: the plankton and the filter-feeder; the virus, mammal, and mosquito.

Mangroves, as Elizabeth Povinelli notes, have no skin around their outer boundary, but within them there are many membranes, many shared cyclings.[23] A mangrove forms a thick, slow membrane, and inside it, membranes inside membranes, each one responsively written and writing within chiasmatic processes, (as was described for an oyster in the first chapters). Membraning processes are abundant and varied in the muddy environment, sorting and resorting distinctions, and homogenising much of the distinction back into the generality of mud.[24] It is attentive work; careful in the sense Neimanis suggests when she describes the communication involved through these in-betweens. All these membranes slow and filter. If we speed up the flow, disallowing pooling and removing membranes, send everything down the gutter, we diminish the attention to the communicative processes of differentiating and decomposing that complicate the world. The desire and functionality of increasing efficiency exemplified in the building of embankments and dredging speeds up flow and destroys the capacity of slowing and communicating that enables diversity.

Within its increasingly neat suburban setting, the Cooks River mangroves are unsettling, not because they threaten or destroy, but because they continue to suffer and accommodate, continue to propose more generous alternatives, continue finding footholds and generating nourishing environments along the margins of the human-exceptionalist geography. They continue to express a wildness upon the Sydney urban geography in the form of a muddy, generative, and insistent mangrove imagination that annoys the neater, human-focussed, urban imaginaries. That reminder beckons as an otherworld of radically different stories, lived alongside, reading and writing its estuarine narrative with an alternative semiotic material grammar. In this way, the mangrove-edged embankments of the Cooks River write a form of dwelling as resistance upon the urban geography.[25] Mangrove resistance challenges the anthropocentric, the human exceptionalist, and the colonial settler geographies of the city. It reminds the landscape of its other possibilities.

Mud and tragedy

Mud

It is not surprising to consider mud together with walls; mixed with straw in adobe it has been part of human wall building for thousands of years. Despite its apparent abundance, mud is not some elemental always-there. It becomes from the combinations of biological and geological processes and products. Mud began accumulating on land after plants appeared. Before terrestrial plants, rivers ran over rock and gravel. They developed more defined channels as plants and decaying plant matter began building up mud banks.[26] In the mangal, much of the characteristics of mangrove ways of walling also relate to the qualities of their mud. There are important mangrove environments with exceptional relationships with mud. Mangroves living on coral atolls, for example, establish on nutrient poor coral sands and slowly build up mud from their own leaf litter and other detritus. As I push my fingers into the mud among the mangroves the fine sliminess and slight greenish tinge surprise me. This mud is teeming with bacterial, algal, and fungal life.[27] My fingers also encounter masses of matted fibres; fine roots, mycorrhiza, and what seems like human hair. Pieces of broken and rotting pneumatophore and leaf, building gravel, and plastics are here as well, but the general feeling is of a fine and surprisingly even consistency. Like some tactile, material virtuality, mud is always in suspension and movement to some degree, shifted by every changing tide yet still providing the ground and remaining itself barely thinkable.

Estuarine mud occupies the lowest position in the catchment, perhaps also in the hierarchy of Sydney's inhabitants, comprised of those who have been broken down as far as they can go, into the smallest specs of stuff washing finally into river mud, mixed with plastics and poisons in some seldom-noticed margin, the basest spaces of the city. Linear time is undone inside urban river mud. The city's past is democratised here as its discarded things accumulate. It is the city undone and reorganised according to the mud's own rhythms, sorted by tide, flood, gravity, and life's movements. It is an organisational ordering beyond anthropologic, a mangaling, beyond any singular controlling logic.

Everything that the mud welcomes becomes, in a sense, one mud; one large sucking sponge, pulsing with the tide, mixed with diverse living beings. But this view is also not the case. Mud is never a one. Every area of mud is entirely different from every other bit. No piece of mud has the same composition or history as any other part. Being in mud is to be particularly situated. Steven Connor suggests mud is:

> a kind of slack, exhausted, overfolded dough, a material in which all possible lines of folding have been included, to the point where there is no longer any difference or potential left. Like many jokes, it embodies a double movement, whereby a signal is first degraded into noise, but then the noise rises up, like a tar baby or creature formed out of mud, in the form of a new signal.[28]

Just as Connor describes here, the Cooks River mud seems "overfolded", as though the semiosis of the mud has become too densely folded to be meaningful as it once was. The folds of semiotic materiality that write difference and relationality into all the forms and activities of the catchment are here all "exhausted". But from it the new arises, as mangrove, oyster, crab, heron, and mosquito. These bodies made of the catchment undone and redone, refolded with the ocean into new forms through the tidal and filter-feeding grammars of the mangroves.

Muddy opacity and resistance gift delay, risk, and opportunity into a material semiotic narrative. Within the spongy mud, flow is not stopped, but is slowed, facilitating flocculation and settling. Tidal and stream flows then provide randomising stirring-up of this accumulated sediment, and the opacity of the mud enables both opportunity and chancy encounter. Plankton, food waste, and petrochemicals come up against filter-feeder, fish comes up against cormorant, bacterium meets dead flesh. The muddiness beckons with its multiple potentials for material narrativising: to lurk as a stalker, to hide as prey, to encounter a companion, to hunt, to filter, to absorb, to move around, to be shaken with potential-increasing randomising, to settle, to hold fast and have the world pass through. Sites of becoming-meaningful are opaque because the constraints that make them possible are folded within them, as membranes of affective figuring-between. Muddy opacity folds risk within the mangrove, so organisms stay alert within obligated orientations. A combination of high levels of both disturbance and opportunity drives mangal adaptability and their remarkable resilience.[29]

It seems impossible to imagine a world without mud, but our bright wealthy cities are already approaching that vision. It is almost as though "eradicate the mud", separate the wet from the dry, is the mantra of human development – so much effort to clean, to sweep, so much good plumbing, and so many machines to wash and vacuum and flush. There are serious predictions that mangroves, as biodiverse ecosystems, will not survive the 21st century[30]. One of the main contributors to this gloomy forecast is that mangroves are running out of mud. Growing as they do in the tidal zone, mangroves are already affected by sea level rise and are doing what they can in response. They have three options. Either they build up higher mud banks, or they move inland, or they die. Building up mud banks occurs as mangrove roots dissipate wave action and allow sediment to settle from turbulence. As this occurs, new mangrove shoots establish on the settling mud, so the mud builds higher. However, in many places, it seems unlikely that mangroves can build up fast enough to manage the predicted sea level rise because of the impacts of dams and sea walls.[31] Most rivers now are dammed so sediment is not arriving where the mangroves need it, and sea walls reduce longshore sediment movement to the estuary areas, so mangroves in many areas are starved of mud. Without sufficient mud, they cannot spread or keep up with sea level rise. While mangroves along the south eastern coast of New South Wales (NSW), including those of the Cooks River, are not thought to be in danger of running out of mud, and are increasing in extent,[32] those in many other areas are more adversely affected, so that in many areas where thick mangal zones might be able to provide protection, this is not likely to be possible.[33] It is particularly notable that even mud, that least of all distinguished beings, the lowest, the excluded, even mud requires care, respect, and attention so it can work its radical transformations.

Tragedy

Jacques Derrida proposes a reading of difference that enfolds it with tragedy. There is tragedy in yearning for closure in an interconnected world, in building walls that will be harmful, temporary, wrongly placed, and insufficient to the task. Indeed, Derrida suggests, the task asked of such walls will remain uncertain and potentially detrimental, despite the call to build them being undeniable. Intimated in Derrida's visualisation; "Sovereignty is round; it is a rounding off", is the creativity, violence and tragedy of processes of working narratives towards closure.[34] Tragic downfall is embedded within a storying of being whose terms of evaluation are established through an assessment of completion, closure or purity. Successful closure would mean the running down of all change, the predictability of relationship, the end of emergence. It is therefore with tragedy, with the inevitable failure of that desired closure, that hope is to be found.

Derrida develops his sense of the tragic within difference in his concepts of the pharmakon and autoimmunity. Tragedy is inevitable within these concepts and is both the way out of and the way into any kind of movement. Autoimmunity as Derrida describes it, loosely relates to the autoimmunity of biological processes,[35] whereby the systems within our bodies that mobilise to protect what is commonly conceived as corporeal integrity from threats, apparently misread or exaggerate their cues and attack the body itself (this understanding of autoimmunity, which ignores the intra-responsivity among the diverse participants involved, is critiqued in Chapter 6). For Derrida, it tells us that in facing an aporia, assessments are being made without surety. Whether these are the assessments being made among cells and microorganisms as they encounter each other within the body, or the processes of democratic decision-making among individuals making their uncertain stake in the future of the policies of their state, the method matters, but no method can negate the aporia and the need to leap into uncertainty. The body-to-come, or the democracy-to-come, is and will continue to be, held open. We, at all our scales, the self who decides which box to tick at an election, the broader public discourse buzzing around an issue, the lymphocyte facing a new potential virus, the immune system taken as a whole, "we" never know clearly what body it is whose future our decisions are contributing to. We are taking risks that may be, in Derrida's sense, autoimmune threats, or they may increase immunity or resilience, but even in this sense, the "of what?" is being negotiated through those very decisions.[36] Matthias Fritsch provides a succinct exposition:

> If self-protection does not in advance know what the self is and what or who the enemy is, but comes to constitute it only in repeated acts

> of protection, and if, further, the self consists in relation to its outside, its environment, and other entities, then the protection will necessarily direct itself against the living self as well.[37]

Immunity and autoimmunity, from this view, become conflated, both perceived outcomes of ecologies of practices among the inhabitants within what might be thought of as an open, experimental field (Haraway's very similar account of immunity is discussed in Chapter 6). Processes of entangled relational becoming (which we can call ecological) require processes that both continue to seek and fail to achieve, closure. Tragedy, such as that encountered as the self-destructive diseases of autoimmunity, is thus a strange saviour. Derrida implies we must love our tragedy, and learn to be sensitive to its warning, for it saves us from the greater death,[38] even as it kills us.

While the walls of human exceptionalism and difference that interested Derrida leave us in an impossible tragedy, mangal walls suggest the potential to avoid this through cosmoecological strategies. Within the thickly knotted inter-affectivities occurring among the mangroves, the walls that come into being are places of ongoing figuring between. They are membranes, in the sense discussed by Neimanis; sites of differentiation where the particular concern is not the creation and maintenance of difference, but the ways difference becomes materially meaningful as it is storied through responsive, communicative, shared cyclings. Humans are not outside these membranes and acknowledging this can shift the ways we consider embankments and other walls.

Host/guest/mangrove

Host/guest

Derrida and Michel Serres both develop their thinking about decision and responsibility through working with concepts related to hospitality. They both begin from the premise that it is only once a space is claimed, the walls built, the inside emptied out, and the host's sovereignty established upon the domain, that generous hospitality can begin. The host can invite the chosen guests into the space.[39] Guests are selected from all those others who are excluded outside. Derrida puts it this way: "No hospitality [...] without sovereignty of oneself over one's home, but since there is also no hospitality without finitude, sovereignty can only be exercised by filtering, choosing, and thus by excluding and doing violence".[40] Within the human context, pleasing the potential human host, (who might be future employer, patron, friend, or neighbour) then becomes another familiar normative social mechanism that walls bring to bear upon society. From the perspective of the host, normative pressure comes from the kudos hospitality offers, through signalling the good fit that has been achieved with the value-defining walls of the society to have the capacity for hospitality. Derrida, very aware of this violence, turns to face it as the call to responsibility: "the problem of hospitality was coextensive with the ethical problem. It is always about answering for a dwelling place, for one's identity, one's space, one's limits, for the ethos as abode, habitation, house".[41] Solid walls need to be answered for if their violences are not to be denied. In a mangal context, where walls are membranous, where ecological interdependencies and co-becomings muddy the categories of host and guest, this burden of responsibility shifts. The decision of the membrane is open to participatory negotiation; it is an ongoing responsive work of co-becoming.

Is hospitality possible without the violence and cleavage that writes host as inside and guest as invited in from outside, always some kind of exception? If this violence is indeed necessary for hospitality, as Derrida's exploration suggests, the challenge is either to fully face its consequences and come to terms with what ethics entails within this impossible space, or to consider how to designify, disempower, or remove the walls that enable this violence and this hospitality. These are challenges Derrida embarks upon with his discussion of unconditional hospitality.[42] This is hospitality that denies itself the power of sovereignty, and welcomes without discernment of difference, without conditions, without decision. This host must, despite maintaining the walls of her territory, be powerless in the face of any guest who seeks to enter, must deny the right to except, be only open and refuse to deny, be host who is no longer host. Derrida offers unconditional hospitality as a provocation, not to suggest the impossible is possible, but to demonstrate how far from ethically unproblematic gestures of conditional hospitality are, to prise open a greater depth of accounting from those who may assume that conditional hospitality is ethically adequate.

As Serres' points out, host and guest are both hosts and both guests, but which role they perform differs with the relational frame and the issue of concern. It may not be so easy to tell who is in which role. The mangroves planted or invited into eco-designed embankment areas return many benefits to their human hosts, hosting them in terms of generously offered services the humans are invited to enjoy. "Thus, offering someone a glass of water, or a bed for the night, is or is not hospitality depending entirely on the relation between the one offering and the other accepting or refusing", Judith Still writes, reflecting this complex relational aspect of hospitality.[43] Serres elaborates upon this tense interdependency by moving his choice of words from "gift" to "given" to emphasise the scale of the legacy of planetary benevolence humans have inherited as guests/parasites and laid to waste by mistaking themselves as hosts: "What can we render to the world that gives us the given, the totality of the gift? [...] The balanced answer would be: the totality of our essence, reason itself".[44] He pleads for human rationality, as gifted, to be put to work to redress the error.

Serres' relational and contextual descriptions of hospitality are helpful in addressing a problem with conceptualisations, including Derrida's, which maintain a commitment to the significance of walls as foundational for hospitality and ethics. Derrida remains committed to thinking from the perspective of the sovereign host, the insider with the potential for generosity, seeking to understand their own responsibilities as host towards the other as guest, or from the perspective of the outsider who would wish to be inside or who grapples with indebtedness. But there are other perspectives from which to think this issue. In particular, there are situations where many beings are engaging, each with their capacity to offer generosity in some form, and each with needs and desires that can be met to the point where one simple framing, that of singular host and singular guest, may be an impediment to understanding.

Michael Marder explores this issue within Derrida's thought through a discussion of what he identifies as Derrida's refusal to engage with ecological thinking.[45] He suggests that refocusing Derrida's insights to incorporate an ecological mode of thinking could open up what he identifies as a much-needed avenue of escape from impasses that are the result of remaining within exchange-based models that overvalue boundaries. In constructing his analysis around the relationship between economic and ecological framings of the oikos, the dwelling space, Marder's discussion highlights the significance of walls in the difference between these two framings:

> Whereas ecology opens up the dwelling by articulating the parts of the oikos in a mode different from and deferring the imposition of the law, economy signals the closure of the dwelling through its total determination and invalidation of ecological concerns. The event, however, spans both sides of the unequal equation.[46]

Rather than denying the economic, which he suggests closes walls, Marder wishes to partner it with the ecological, which opens them. It is through such a partnership, the uneasy negotiation of both territory and selves among interdependent co-inhabitants, that Marder finds generative possibilities for modes of living that can support greater flourishing. His proposal sounds modest, but it is not undemanding because ecological thinking, in his view, is lacking not only within Derrida's work, but much more widely. A similar argument is made by Julian Yates, in the context of the different ways the concept of the gift is developed by Derrida and by Serres. Yates writes: "In thinking the gift, then, Derrida does not aim to exit the circle of exchange. [...] For Serres, however, the gift becomes a given, a closed, self-regulating system".[47]

Derrida's interest, however, is to challenge those with walls and sovereign power to consider their responsibilities. He is speaking *within* the context of existent walls and calling out the responsibility their violences demand. That work still has its currency. Serres engagement is from a somewhat more ecological ontology, where the sovereignty built into the walls is from the beginning already being undone by parasites. As mangrove roots and salty water undermine the embankments, subject and object, host and parasite are relative, uncertain categories. "By the end of his reading, no differences will remain between rats, humans, microbes, breaks in the message, static. Ontology collapses. Hospitality reigns", he writes.[48] Serres more ecological reading does not deny or replace Derrida's, rather it offers vocabularies and visions for more ecological, membranous futures.

Mangrove hospitality

Compared with the aims of human estuary engineering, mangroves aim for a muddier, changeable environment, open to diverse creatures who in turn contribute to fertility and adaptiveness. Mangroves are generously hospitable, welcoming so many to join their assemblage. Might Derrida's unconditional hospitality fit as a description of mangal hospitality? Mangroves invite into their realm their own selection of companion species in what might seem to be unconditional, unwalled hospitality, but to make this interpretation would be to ignore the complexity of conditions and relations. Decision-processes are occurring within chemical, physical, cultural, cognitive, and semiotic frames, and these overlap. Every one of those decision-processes ripples across the options and opportunities of others. Different beings are becoming situated as host or guests differently within each varying frame of spatial and temporal reference. Once we step into an ecological semiosis, the roles of host and guest flicker on and off and swap around between each other so rapidly. The sovereign host and the singular invited guest are concepts that remain bound to assumptions of simple exchange. As Marder points out, they do not transfer easily into the complexities and multiple shifting directionalities of ecological thinking.[49] On the other hand, they are relevant to understanding and deconstructing the motivations and sustaining narratives of anthropocentrism.

There *are* walls that have written the simplicity of human exceptionalism upon the landscape, and so much human decision-making presupposes the human exceptionalism and sovereignty that the walls convey. The Cooks River walls continue to maintain and reflect the values of the legacy of colonial sovereignty. They continue to control the water of the region towards those human uses established through the colonial enterprise and away from others, a control visibly challenged by the river's mangroves. Derrida's call to consider unconditionality is contextually appropriate within this human exceptionalist and exemptionalist landscape and offers a means to challenge some of its assumptions. Serres' thinking with parasitism seems to be operating as a step between the economic and the ecological in this regard. He is playing between an economic model and an ecological one, tossing the two together in his writing practice. What comes through this exercise is the parasite, which flits and flips between the economic thinking of host and guest, and the ecological thinking of processes within relations, a thinking that continually redraws boundaries and swaps perspective across each emerging boundary.

Derrida's challenge asked us to imagine together both walls and the refusal to use the sovereign power they provide. It was a challenge to those who are, now, already holders of that power, and it was a challenge that was well aware of its own impossibility. Mangrove hospitality seems to offer a model without walls or sovereign, and clearly one that is possible. But to say there are no walls denies the tensions and agonisms that are likely at play there as entities jostle and negotiate their obligations and opportunities. Perhaps it is rather the case that any walls or skins that are able to appear – as protection, to hold off the parasites, to reduce the competition – must be particularly attended to, lest they remove the vulnerability that each entity requires in order to remind it of its ongoing obligations within the emergent being that is the host of all. Perhaps this is a useful way to interpret Derrida's call to think towards impossible unconditionality, to take it as a warning. Should I find myself the beneficiary of a skin or wall, or with the desire to erect a wall to save myself the effort of other methods of sustaining myself within an irritating environment, beware. My comfort may be at the expense of something far more precious, some larger host, who may turn its back if its guests/parasites remove their attentive care (as described in Chapter 4 with Povinelli's account of the mangrove estuary Tjipel). Or should I be the witness of such an action by another, consider my responsibility as guardian and guest of that larger host. Serres suggests it is indeed the appropriate place of such skins and walls to be attacked as opportunities by parasites.

Serres asks us to leave our walled security, and come out into attentive, frictional, unpredictable relationships. Without walls to keep the parasites away, we have to attend, negotiate, placate, distract, fight, persuade, collaborate, care – we need a wide range of relationship strategies to figure trajectories among all the others. For Serres it is only this kind of alert, adaptive, wily engagement that counts as living. Security inside the walls of sovereignty is a kind of death in his assessment. In *The Natural Contract*, he suggests the solution to the tragic separation that has come from securing ourselves "lies in placing one's defences and security outside the body: in relations".[50] Again, the call is to seek ways of being with others that recognise the value of being without skin or wall, that require our knowledge that because we are all vulnerable together, we have no better option than to focus our attention on the well-being of all our hosts and all our guests and parasites, and the ways we exchange these roles amongst ourselves. This is the kind of vulnerability introduced by Judith Butler as "one of the most important resources from which we must take our bearings and find our way",[51] and which Erinn Gilson describes as "the basis for learning and for empathy, connection, and community".[52] Where Derrida provides the conceptual apparatus for deconstructing and revealing the points of leverage within the processes of desiring security, Serres offers some ropes to grasp[53] to guide a journey elsewhere. I note, however, that Serres' account at times seems to forget the recognition he shows elsewhere that ongoing tensions between trends towards security and those towards relations will remain problematic.

Mangroves and humans pull in opposite directions when it comes to estuarine design. It is tempting to read a story here of competing approaches to worlding estuaries, two types of directionality, parallel versions of time, two ways to inhabit an edge. It may then appear the colonial drive to wall edges has in some perverted sense won the competition, by defining the terms in its favour, and by dominating the landscape visually and functionally. This reading makes it hard to adequately value the cooperation that occurs between these communities, the knots that entangle the stories. Mangroves are finding their way around and into the embankment walls, muddying the engineered categories of nature and history, originality and authenticity, muddying economic values and aspirations. We thought we were making money by destroying them, but it turns out they are of extraordinary economic value. Similarly, previous attitudes that regarded mangroves as aesthetically ugly health hazards are reversing. On the Cooks River, as in many parts of the world, mangroves are rapidly gathering human friends.

Riverbank negotiations

The Cooks River mangroves have accommodated much from the human world into their community, including toxins and refuse. They have gone further and gathered humans to help them in weekend working bees. "The Mudcrabs" is a sizeable and committed group of river helpers who pick up rubbish and plant the banks of the river to help it regain some degree of ecological complexity.[54] Mangrove needs are also being considered in new embankment designs. Sections of embankment along the Cooks River have been demolished and rebuilt with coarse stone blocks, sometimes interspersed with carefully located and well-bounded terraces where mud can accumulate to accommodate mangroves and saltmarsh in controlled areas.[55] Humans are hosts here, opening spaces in their walls and inviting in mangroves and saltmarsh. These eco-design projects are generally very popular, supported by scientists, local groups, residents, businesses, and councils. They are successful in offering support for individuals and communities of local plant species, they are affordable compared with other engineering alternatives, they are integrated to be aesthetically pleasing within the urban landscape, and they are located to fit in conveniently with other human-oriented functions of the waterside. They often sit alongside running paths, or picnic and sports facilities. Several of them are regarded as educational sites, safe places for children and curious adults to learn more about marine and estuarine life and the innovative eco-design features.

Mangroves are indeed using these sites, although they are constrained within small compartments within the redesigned embankments. Their human hosts have engineered strict controls. This is highly conditional hospitality,[56] but nonetheless these eco-designed embankment environments can be read as some kind of meeting space, some space where semiotic material conversations about multispecies catchment design can be explored.[57] These eco-designs demonstrate some willingness on the part of humans to be polite (in Vinciane Despret's sense, as discussed in Chapter 6) in negotiating riverbank landscaping with mangroves. They demonstrate a basic concern for mangrove needs, and invite mangrove life, but they limit mangroves capabilities. Within these sites, mangroves are constrained, controlled, and unable to build up complex multispecies worlds. These are somewhat domesticated mangroves, invited to live, but only with diminished agency.

These sites demonstrate some of the significant challenges of negotiating futures as participants within more-than-human-worlds from positions of power. Povinelli considers this question in relation to the future of the Northern Australian mangrove estuary, Tjipel, which featured in Chapter 4. As she considers the challenges of trying to include Tjipel and other geolontological beings into shared semiotic futures, she reflects upon the troubling effect of material semiotic power. "The generosity of extending our form of semiosis to them forecloses the possibility of them provincializing us" she writes, aptly capturing the closures and exclusions practiced by acts of semiotic sovereign hospitality.[58] She argues that because we are powerful enough to foreclose negotiations with others, if we are to avoid the violences caused by the imposition of anthropocentric rationality upon other ways of being meaningful, we must make efforts to allow our logic to be "decentred by noise".[59] By this, she means we must actively seek narrative strategies to assist us to restrain our logic, to resist neatly writing a happy ending to shared negotiations, and to remain open to a shared material semiosis whose meanings must remain only partially accessible. Chapter 6 presented and discussed a selection of material semiotic narrative strategies that respond to that proposal. These approaches come to the fore when difference is understood as actively negotiated within relations among diverse meaning-makers.

The mangal marks difference, the difference between land and water, salt, and fresh, by opening the space to a thickness of being-together in turbulent cooperative, competitive, and agonistic negotiations between multitudes. It shows that there are ways of territorialising together and apart that do not place a sovereign host at the gate or embed power into a non-negotiable future or write it into a non-negotiable geography. These mangal ways are challenging the legacy of colonial values inscribed upon the suburban geography and water economy by the embankments of the Cooks River. Mangrove agency is suggesting multispecies futures for the city. Eco-designed embankments still function to maintain the river's edges, but are accommodating, held together by living beings, relying upon many species and individuals choosing to move in and settle within them. The human designers planned porous, accommodating, adaptive boundaries that mangroves found inviting. Plants and designers are putting some relationship work back into the process of negotiating these troubled territories. The relations are far from equal, far from achieving the degree of "good manners" to which Despret's ideal of more-than-human politeness aspires, but they are much less ill-mannered than the stark embankments.

The Cooks River mangroves offer a form of resistance to the settler colonial writing of the catchment by living as an "other alongside". This mangal other alongside finds difference negotiated in relations. It suggests muddy, unsettled, and risky edges continually in flux. Derrida's tragic impossibility might be an apt place to end if capitalist walls continue to divide and establish difference as essential. But mangal walls offer alternatives. They are made from inter-folded diverse narratives of many different organisms coming together and attending responsively to each other, they write difference thickly and complexly within relational processes that mix up host and guest, hospitality and parasitism. Mangroves offer a more problematic, messier hospitality that displaces the sovereign from the gate by folding sovereignty in among many inter-affective stories. Their occupancy of the riverbanks right alongside the hard edges of the Cooks River embankments suggests more inclusive alternatives to differencing for those of us aspiring towards generous ways of living more-than-human futures.

Notes

1 Michel Serres, *The Natural Contract*, trans. Elizabeth MacArthur and William Paulson (Ann Arbor: The University of Michigan Press, 1995), 2.

2 Daniel M. Alongi, "Mangrove Forests: Resilience, Protection from Tsunamis, and Responses to Global Climate Change," *Estuarine, Coastal and Shelf Science* 76, no. 1 (2008): 1–13; Farid Dahdouh-Guebas, Loku Jayatissa, Diana Di Nitto, Jared Bosire, et al., "How Effective Were Mangroves as a Defence against the Recent Tsunami?", *Current Biology* 15, no. 12 (2005): R443–7; Kandasamy Kathiresan and Narayanasamy Rajendran, "Coastal Mangrove Forests Mitigated Tsunami," *Estuarine, Coastal and Shelf Science* 65, no. 3 (2005): 601–06.

3 Environmental Justice Foundation, *Mangroves: Nature's Defence against Tsunamis—A Report on the Impact of Mangrove Loss and Shrimp Farm Development on Coastal Defences* (London: Environmental Justice Foundation, 2006), http://ejfoundation.org/report/mangroves-natures-defence-against-tsunamis

4 Alongi, "Mangrove Resilience."

5 As a result of this, coastal mangroves are often damaged by storms. The Cooks and Georges River mangroves are estuarine, located up-river from the coast in bays well protected from many storms. Most of the mangroves that protected communities during the 2004 tsunami were also themselves damaged or destroyed, an estimated 30,000 hectares, as noted by Fred Pearce, "A Decade

after Asian Tsunami New Forests Protect the Coast," *Yale Environment 360* (December 2014), https://e360.yale.edu/features/a_decade_after_asian_tsunami_new_forests_protect_the_coast

6 Ken W. Krauss et al., "How Mangrove Forests Adjust to Rising Sea Level," *New Phytologist* 202, no. 1 (2014), https://dx.doi.org/10.1111/nph.12605

7 Campaigns to replant mangroves in areas where they had been destroyed by the tsunami or by shrimp farming were largely unsuccessful at first, due to poor planning and lack of knowledge but expertise quickly developed and the projects are experiencing greater success. See Pearce, "Decade."

8 Ivan Valiela, Jennifer L. Bowen, and Joanna K. York, "Mangrove Forests: One of the World's Threatened Major Tropical Environments," *BioScience* 51, no. 10 (2001), https://dx.doi.org/10.1641/0006-3568

9 Valiela et al. "Mangrove Forests" : 810.

10 Cooks River Alliance, *Cooks River Alliance Annual Report 2018–19*, 2019, https://cooksriver.org.au/wp-content/uploads/2019/09/CRA-Annual-Report_2018-19.pdf; Georges Riverkeeper, "Georges River Pollution", 2020, accessed 5 August 2020, https://georgesriver.org.au/learn-about-the-river/georges-river-pollution

11 Norman Duke and Klaus Schmitt, "Mangroves: Unusual Forests at the Seas Edge," in *Tropical Forestry Handbook*, ed. Laslo Pancel and Michael Kohl (Berlin: Springer-Verlag, 2015): 1693.

12 Pearce, "Decade."

13 David Probandt et al., "Microbial Life on a Sand Grain: From Bulk Sediment to Single Grains," *The ISME Journal* 12 (2017), https://dx.doi.org/10.1038/ismej.2017.197

14 The process of oyster filter-feeding is described in detail in the first part of this book, "Strange Reflexivities".

15 Kandasamy Kathiresan and Brian Lynn Bingham. "Biology of Mangroves and Mangrove Ecosystems," *Advances in Marine Biology* 40 (2001): 27–36.

16 Pentti Määtänen, "Space, Time, and Interpretation," *Place and Location: Studies in Environmental Aesthetics* 5, no. 1 (2006): 15, www.eki.ee/km/place/pdf/kp5_01_maattanen.pdf

17 Michel Serres expresses this, writing:

> Let's not claim we alone write. Oil and water don't mix; bodies select partners for their combinations to the exclusion of other elements; crystals endowed with impurities rectify the direction of certain flows. The act of choosing doesn't concern us alone. Ice sheets, cliffs, radioactive bodies engram memories. Let's not claim we alone remember. In short, the things themselves, inert, as well as the living things exchange elements, energy and information, preserve this latter, spread it, select it. Let's not claim we alone devote ourselves to exchange. This writing, these decisions, these memory storages, these codings, among other examples, endow objects with quasi cognitive properties. "It thinks" in the sense of "it rains" exists as much as "I think" or "we think",

in *The Incandescent*, trans. Randolph Burks (London: Bloomsbury Academic, 2018), 190–91, Kindle.

18 "We conceive social space by attaching meanings to perceived physical sign-vehicles, and these meanings exist as habits of social practice", writes Määtänen in, "Space", 15.

19 Mark Kinver, "'Let Mangroves Recover' to Protect Coasts," *BBC News Science and Environment*, 19 September 2016, www.bbc.com/news/science-environment-37386267

20 Murray Trembath, "Extensive Damage to Beaches and Retaining Walls around Bate Bay from Extreme Ocean and Weather Conditions," *St George & Sutherland Shire Leader*, 30 July 2020, accessed 10 September 2020, www.theleader.com.au/story/6856779/big-seas-undermine-the-wall-at-north-cronulla/

21 Kinver, "'Let Mangroves Recover' to Protect Coasts"; Russell Shorto, "How to Think Like the Dutch in a Post-Sandy World," *The New York Times Magazine*, 9 April 2014, accessed 5 August 2020, www.nytimes.com/2014/04/13/magazine/how-to-think-like-the-dutch-in-a-post-sandy-world.html

22 Astrida Neimanis, "Thinking with Matter, Rethinking Irigaray: A 'Liquid Ground' for Planetary Feminism," in *Feminist Philosophies of Life*, ed. Chloe Taylor and Hasana Sharp (Montreal and Kingston: McGill-Queen's University Press, 2016), 59.

23 Elizabeth A. Povinelli, *Geontologies: A Requium to Late Liberalism* (Durham and London: Duke University Press, 2016), chapter 4, paragraphs 14–15, Kindle; Elizabeth A. Povinelli, "Transgender Creeks and the Three Figures of Power in Late Liberalism," *Differences: A Journal of FeministCultural Studies* 26, no. 1 (2015): 184, https://dx.doi.org/10.1215/10407391-2880645

24 Stephanie Fishel urges a reading of state borders as membranous spaces actively involved in making difference, not just inert barriers, in *The Microbial State: Global Thriving and the Body Politic* (Minneapolis: University of Minnesota Press, 2017), Kindle.

25 Nicholas Ferguson, "Dwelling as Resistance," *Places Journal* (2019), https://placesjournal.org/article/dwelling-as-resistance/

26 Laura Poppick, "The Origin of Mud," *Knowable Magazine*, 17 August 2020, www.knowablemagazine.org/article/living-world/2020/origin-mud

27 Kathiresan and Bingham, "Biology of Mangroves."

28 Steven Connor, "Michel Serres's Milieux", *Steven Connor*, 2002, accessed 18 August 2020, www.stevenconnor.com/milieux/

29 Alongi, "Mangrove Resilience."

30 Beth A. Polidoro, Kent E. Carpenter, Lorna Collins, Norman C. Duke, et al. "The Loss of Species: Mangrove Extinction Risk and Geographic Areas of Global Concern." *PLoS One* 5, no. 4 (2010).

31 Catherine E. Lovelock et al., "The Vulnerability of Indo-Pacific Mangrove Forests to Sea-Level Rise," *Nature* 526 (2015), https://dx.doi.org/10.1038/nature15538; Helen Scales, "Mangrove Forests Are Being Starved of Mud," *Hakai Magazine*, 12 November 2015, www.hakaimagazine.com/news/mangrove-forests-are-being-starved-mud/; Eric Gilman et al., "Threats to Mangroves from Climate Change and Adaptation Options: A Review," *Aquatic Botany* 89, no. 2 (2008), https://doi.org/10.1016/j.aquabot.2007.12.009

32 Lovelock et al., "Vulnerability"; Tim Low, *Climate Change and Terrestrial Biodiversity in Queensland* (Brisbane: Department of Environment and Resource Management, Queensland Government, 2011).

33 Manon Besset et al., "Mangroves and Shoreline Erosion in the Mekong River Delta, Viet Nam," *Estuarine, Coastal and Shelf Science* 226 (2019), https://doi.org/10.1016/j.ecss.2019.106263

34 Jacques Derrida, *Rogues: Two Essays on Reason*. Translated by Pascal-Anne Brault and Michael Naas. Stanford: Stanford University Press, 2005, 13.
35 Andrea Timár explains how Derrida's version of autoimmunity is significantly different from the biological process, in "Derrida and the Immune System/ Terror(Ism) and Aesthetics", *Et al. Critical Theory Online* (2015), http://etal.hu/en/archive/terrorism-and-aesthetics-2015/derrida-and-the-immune-system/
36 Derrida, *Rogues*, 52.
37 Matthias Fritsch, "An Eco-Deconstructive Account of the Emergence of Normativity in 'Nature'," in *Eco-Deconstruction: Derrida and Environmental Philosophy*, eds. Matthias Fritsch, Philippe Lynes, and David Wood, (New York: Fordham University Press, 2018): 289.
38 Derrida writes, for example, of autoimmunity as robing suicide itself of its meaning, because of its threat to ipseity, in *Rogues*, 45.
39 See Chapter 8 for a discussion of the hospitality of the colonial geography of the Cooks River embankments in conditionally welcoming back mangroves.
40 Jacques Derrida and Anne Dufourmantelle, *Of Hospitality: Anne Dufourmantelle Invites Jacques Derrida to Respond* (Stanford: Stanford University Press, 2000), 55.
41 Derrida and Dufourmantelle, *Of Hospitality*, 149.
42 Ibid.
43 Judith Still, *Derrida and Hospitality: Theory and Practice* (Edinburgh: Edinburgh University Press, 2010), 20.
44 Serres *Natural*, 90.
45 Michael Marder, "Ecology as Event," in *Eco-Deconstruction: Derrida and Environmental Philosophy*, eds. Matthias Fritsch, Philippe Lynes, and David Wood (New York: Fordham University Press, 2018): 141–164.
46 Marder, "Ecology", 160–161.
47 Julian Yates, "'The Gift Is a Given': On the Errant Ethic of Michel Serres," in *Mapping Michel Serres*, ed. Niran Abbas (Ann Arbor: The University of Michigan Press, 2005), Same and Other, paragraph 13, Kindle.
48 Yates, "The Gift Is a Given", From Parasite to Symbiont, paragraph 3.
49 See Marder, "Ecology".
50 Serres, *Natural*, 104.
51 Judith Butler, *Precarious Life: The Power of Mourning and Violence* (London and New York: Verso, 2004), 30.
52 Erinn Cunliffe Gilson, *The Ethics of Vulnerability: A Feminist Analysis of Social Life and Practice* (New York: Routledge, 2014), 2.
53 Serres, *Natural*, 104–11.
54 Cooks River Valley Association, "The Mudcrabs: Cooks River Eco Volunteers," accessed 11 July 2018, www.crva.org.au/?page_id=7
55 Eco Logical Australia Pty Ltd, *Cooks River Estuarine Vegetation Management Plan* (City of Canterbury, Marrickville and Rockdale Councils, 2010).
56 Jacques Derrida has called hospitality "hostipitality", to reveal the hostility folded within it, in Jacques Derrida, "Hostipitality", *Angelaki* 5, no. 3 (2000), https://dx.doi.org/10.1080/09697250020034706
57 Thom van Dooren explores numerous examples of how human efforts to open hospitality to non-humans are inevitably partial, conditional and far from ideal. He finds cruelty, violence and tragedy folded into the best that we can do, in

Flight Ways: Life and Loss at the Edge of Extinction (New York: Columbia University Press, 2014).

58 Povinelli, *Geontologies*, 142.

59 Ibid.

References

Alongi, Daniel M. "Mangrove Forests: Resilience, Protection from Tsunamis, and Responses to Global Climate Change." *Estuarine, Coastal and Shelf Science* 76, no. 1 (2008): 1–13. https://dx.doi.org/10.1016/j.ecss.2007.08.024

Besset, Manon, Nicolas Gratiot, Edward J. Anthony, Frédéric Bouchette, et al. "Mangroves and Shoreline Erosion in the Mekong River Delta, Viet Nam." *Estuarine, Coastal and Shelf Science* 226 (2019): 106263. https://doi.org/10.1016/j.ecss.2019.106263

Butler, Judith. *Precarious Life: The Power of Mourning and Violence*. London and New York: Verso, 2004.

Connor, Steven. "Michel Serres's Milieux." *Steven Connor*, 2002. Accessed 18 August 2020. www.stevenconnor.com/milieux/

Cooks River Alliance. *Cooks River Alliance Annual Report 2018–19*. 2019. https://cooksriver.org.au/wp-content/uploads/2019/09/CRA-Annual-Report_2018-19.pdf

Cooks River Valley Association. "The Mudcrabs: Cooks River Eco Volunteers." Accessed 11 July 2018. www.crva.org.au/?page_id=7

Dahdouh-Guebas, Farid, Loku Jayatissa, Diana Di Nitto, Jared Bosire, et al. "How Effective Were Mangroves as a Defence against the Recent Tsunami?", *Current Biology* 15, no. 12 (2005): R443–7. https://dx.doi.org/10.1016/j.cub.2005.06.008

Derrida, Jacques. "Hostipitality." *Angelaki* 5, no. 3 (2000): 3–18. https://dx.doi.org/10.1080/09697250020034706

———. *Rogues: Two Essays on Reason*. Translated by Pascal-Anne Brault and Michael Naas. Stanford: Stanford University Press, 2005. Originally published as *Voyous: Deux Essais sur la Raison,* Éditions Galilée, 2003.

Derrida, Jacques and Anne Dufourmantelle. *Of Hospitality: Anne Dufourmantelle Invites Jacques Derrida to Respond*. Stanford: Stanford University Press, 2000.

Duke, Norman and Klaus Schmitt. "Mangroves: Unusual Forests at the Seas Edge." In *Tropical Forestry Handbook*, edited by Laslo Pancel and Michael Köhl, 1693–1724. Berlin: Springer-Verlag, 2015.

Eco Logical Australia Pty Ltd. *Cooks River Estuarine Vegetation Management Plan*. City of Canterbury, Marrickville and Rockdale Councils, 2010.

Environmental Justice Foundation. *Mangroves: Nature's Defence against Tsunamis—A Report on the Impact of Mangrove Loss and Shrimp Farm Development on Coastal Defences*. Environmental Justice Foundation: London, 2006. http://ejfoundation.org/report/mangroves-natures-defence-against-tsunamis

Ferguson, Nicholas. "Dwelling as Resistance." *Places Journal* (September 2019). https://placesjournal.org/article/dwelling-as-resistance/

Fishel, Stephanie R. *The Microbial State: Global Thriving and the Body Politic*. Minneapolis: University of Minnesota Press, 2017, Kindle.

Fritsch, Matthias. "An Eco-Deconstructive Account of the Emergence of Normativity in 'Nature'." In *Eco-Deconstruction: Derrida and Environmental Philosophy*, edited by Matthias Fritsch, Philippe Lynes, and David Wood, 279–302. New York: Fordham University Press, 2018.

Gilman, Eric, Joanna Ellison, Norman Duke, and Colin Field. "Threats to Mangroves from Climate Change and Adaptation Options: A Review." *Aquatic Botany* 89, no. 2 (2008): 237–50. https://doi.org/10.1016/j.aquabot.2007.12.009

Gilson, Erinn Cunliffe. *The Ethics of Vulnerability: A Feminist Analysis of Social Life and Practice*. New York: Routledge, 2014, Kindle.

Kathiresan, Kandasamy and Brian Lynn Bingham. "Biology of Mangroves and Mangrove Ecosystems." *Advances in Marine Biology* 40 (2001): 81–251.

Kathiresan, Kandasamy and Narayanasamy Rajendran. "Coastal Mangrove Forests Mitigated Tsunami." *Estuarine, Coastal and Shelf Science* 65, no. 3 (2005): 601–06. https://dx.doi.org/10.1016/j.ecss.2005.06.022

Kinver, Mark. "'Let Mangroves Recover' to Protect Coasts." *BBC News Science and Environment*. 19 September 2016, www.bbc.com/news/science-environment-37386267

Krauss, Ken W., Karen L. McKee, Catherine E. Lovelock, Donald R. Cahoon, et al. "How Mangrove Forests Adjust to Rising Sea Level." *New Phytologist* 202, no. 1 (2014): 19–34. https://dx.doi.org/10.1111/nph.12605

Lovelock, Catherine E., Donald R. Cahoon, Daniel A. Friess, Glenn R. Guntenspergen, et al. "The Vulnerability of Indo-Pacific Mangrove Forests to Sea-Level Rise." *Nature* 526 (2015). https://dx.doi.org/10.1038/nature15538

Low, Tim. *Climate Change and Terrestrial Biodiversity in Queensland*. Brisbane: Department of Environment and Resource Management, Queensland Government, 2011.

Määtänen, Pentti. "Space, Time, and Interpretation." *Place and Location: Studies in Environmental Aesthetics* 5, no. 1 (2006): 11–20. www.eki.ee/km/place/pdf/kp5_01_maattanen.pdf

Marder, Michael. "Ecology as Event." In *Eco-Deconstruction: Derrida and Environmental Philosophy*, edited by Matthias Fritsch, Philippe Lynes, and David Wood, 141–164. New York: Fordham University Press, 2018.

Neimanis, Astrida. "Thinking with Matter, Rethinking Irigaray: A 'Liquid Ground' for Planetary Feminism." In *Feminist Philosophies of Life*, edited by Chloe Taylor and Hasana Sharp, 42–66. Montreal and Kingston: McGill-Queen's University Press, 2016.

Pearce, Fred. "A Decade after Asian Tsunami New Forests Protect the Coast." *Yale Environment 360* (December 2014). https://e360.yale.edu/features/a_decade_after_asian_tsunami_new_forests_protect_the_coast

Polidoro, Beth A., Kent E. Carpenter, Lorna Collins, Norman C. Duke, et al. "The Loss of Species: Mangrove Extinction Risk and Geographic Areas of Global Concern." *PLoS One* 5, no. 4 (2010). https://dx.doi.org/10.1371/journal.pone.0010095

Poppick, Laura. "The Origin of Mud." *Knowable Magazine*. 17 August 2020. www.knowablemagazine.org/article/living-world/2020/origin-mud

Povinelli, Elizabeth A. *Geontologies: A Requium to Late Liberalism*. Durham and London: Duke University Press, 2016, Kindle.

———. "Transgender Creeks and the Three Figures of Power in Late Liberalism." *Differences: A Journal of Feminist Cultural Studies* 26, no. 1 (2015): 168–87. https://dx.doi.org/10.1215/10407391-2880645

Probandt, David, Thilo Eickhorst, Andreas Ellrott, Rudolf Amann, et al. "Microbial Life on a Sand Grain: From Bulk Sediment to Single Grains." *The ISME Journal* 12 (2017): 623. https://dx.doi.org/10.1038/ismej.2017.197

Scales, Helen. "Mangrove Forests Are Being Starved of Mud." *Hakai Magazine*. 12 November 2015. www.hakaimagazine.com/news/mangrove-forests-are-being-starved-mud/

Serres, Michel. *The Incandescent*. Translated by Randolph Burks. London: Bloomsbury Academic, 2018, Kindle. Originally published as *L'Incandescent*, Editions Le Pommier, 2003.

———. *The Natural Contract*. Translated by Elizabeth MacArthur and William Paulson. Ann Arbor: The University of Michigan Press, 1995. Originally published as *Le Contract*, Editions François Bourin, 1992.

Shorto, Russell. "How to Think Like the Dutch in a Post-Sandy World." *The New York Times Magazine*. 9 April 2014. Accessed 5 August 2020, www.nytimes.com/2014/04/13/magazine/how-to-think-like-the-dutch-in-a-post-sandy-world.html

Still, Judith. *Derrida and Hospitality: Theory and Practice*. Edinburgh: Edinburgh University Press, 2010.

Timár, Andrea. "Derrida and the Immune System/Terror(Ism) and Aesthetics." *Et al. Critical Theory Online* (2015). http://etal.hu/en/archive/terrorism-and-aesthetics-2015/derrida-and-the-immune-system/

Valiela, Ivan, Jennifer L. Bowen, and Joanna K. York. "Mangrove Forests: One of the World's Threatened Major Tropical Environments." *BioScience* 51, no. 10 (2001): 807–15. Accessed 18 March 2020. https://dx.doi.org/10.1641/0006-3568

van Dooren, Thom. *Flight Ways: Life and Loss at the Edge of Extinction*. New York: Columbia University Press, 2014, Kindle.

Yates, Julian. "'The Gift Is a Given': On the Errant Ethic of Michel Serres." In *Mapping Michel Serres*, edited by Niran Abbas, 190–209. Ann Arbor: The University of Michigan Press, 2005.

10 Conclusion

My aims through this work have been twofold: first, to explore ways of understanding being in-between and, second, to contribute methodologically and theoretically towards non-anthropocentric modes of enquiry. Both these aims came together within the theoretical model I embraced for this research. Semiotic materialism framed both the concepts I investigated and the approach of the book together within a practice. In this research, mangrove material processes and human language are taken to be equivalent in terms of their meaning-making work. The practices of oysters in filter-feeding that build reefs, oyster bodies, and mangrove mud from tidal rhythms and suspended sediment; the processes of infection and immunity changing my body as mosquito saliva and virus enters and interacts; human and mangrove embankment building, are taken here as semiotic practices of material meaning-making. This approach made it possible for the writing to filter Deleuzian concepts through mud banks, to wash embankments through Derridean concepts and to diffract concepts from Donna Haraway with immunology, without privileging one semiotic configuration over the other.

The book has sought to understand how semiotic material narrative coherence becomes, changes, and ceases; how it is practiced "in the middle" through affects and attentions, through concerns that take each other into account, or fail to do so. Each of my three structuring concepts, reflexivity in becoming, relationality, and difference, is a way of proposing questions about the processes that shape narrative coherence. I could find out about how semiotic material coherence comes into being by asking questions about reflexive becoming. I could explore how it is sustained and changed by asking questions about relationality, and I could investigate the way it becomes fractured, shaped, and destroyed by asking questions about difference.

The three other terms used in my section titles: "Strange", "Monstrous", and "Impossible", relate to the affective significance of each of the three semiotic material processes when they are interpreted within an anthropocentric and human exceptionalist frame of reference. From that perspective, the fractal foldings of both-ways-criss-crossing that I have described as processes of reflexive becoming across scales of semiosis, have been

DOI: 10.4324/9781003286493-14

described as strange when they occur across two levels, and nominated as especially relevant for explaining the uniqueness of human consciousness.[1] From the non-anthropocentric perspective of this project, what seem like strange loops are the many and varied foldings of meaning through matter. Similarly, when encountered from an anthropocentric perspective that assumes human exceptionalism, organisms that emerge from swamps, that suck human blood to feed their offspring, and that insert alien life forms into our bodies have been written as monstrous within popular culture. From the more-than-human perspective of this work, the monstrousness is rewritten as ongoing agonistic co-becoming. Finally, through the last part of the book, I drew upon Jacques Derrida's writings outlining the impossibility of difference, which reads as tragic within a context where stasis, integrity, and control are sought for human categories and identities. In a non-anthropocentric semiotic material reframing, difference becomes activities of differencing, semiotic material practices of negotiating criteria and relations. Impossibility does not disappear, but it is accommodated, folded into processes of becoming meaningful. These three supporting terms, therefore, have been provocations, suggesting questions to ask within each area of enquiry, unravelling as the questions were explored.

The first part of the book, "Strange reflexivities: folding in communicative tidal materialities", comprised an enquiry into interstitiality within material semiotic ontologies. I identified two strands of semiotic material thinking, both of which have become important within diverse areas of theoretical research. One, biosemiotics, is relatively clearly delineated as an approach, developed from an alliance between the traditions of ecosystems science, studies of perception and communication, cybernetics, and Peircean semiotics. The other is situated more comfortably within philosophy and is represented in my project predominantly in the work of Haraway, Gilles Deleuze, Brian Massumi, Derrida, Bruno Latour, John Law, and Karen Barad. The two are similar in that both endorse the unification of semiosis and matter within a meaningful materiality. I argued that despite this similarity, the differences between them are notable, leading to important differences in the ontological, epistemological, and ethical implications of each.

Biosemiotics frequently describes a biological materiality that is meaningful because it is layered. Each layer both creates and is a result of the coming into being, within a reflexive relationship, an interpreter at one level and a meaningful concept located at another level. Within this approach, ecological systems function within a nested hierarchy of levels of material meaningfulness. Some biosemioticians locate human symbolic practices as the pinnacle of a singular evolutionary narrative of increasing semiotic capacity – in Jesper Hoffmeyer's terms, increasing semiotic freedom. The alternative version, developed through Deleuze, Latour, Haraway, and Barad eschews fixed layers or any singular narrative of increasing evolutionary progress in semiotic capacity, although, I suggest, it does not move away from layering altogether, or deny any evolutionary progressions in semiotic capacity. With these approaches, the semiotic relation between meaning interpreter and meaningful concept occurs through a fold in some form of immanence; meaning remains in negotiation; a continual matter of concern along an unsteady interface where immanence and emergence shimmer, rather than being defined only in respect to a move of becoming emergent.

"Strange reflexivities" introduced filter-feeding mangal practices into a conversation with semiotic material theory to explore reflexivity as it functions in the becoming-meaningful of matter. Oyster filter-feeding contributes to the meaningfulness of the mangrove ecosystem of the Georges River. Oysters' bodies are inter-responsive with the flow of tides and the currents formed as tidal flows are constrained by the estuary's topography, the buoyancy of water, and the size and shapes and composition of the suspended particles that pass through their bodies. Oyster bodies are shaped in response to these, and in turn direct the current and sort the detritus by passing it through their bodies so that much of the mud of the estuary has been organised, catalogued, coated, and directed into currents, through the bodies of oysters. In making these differences within the estuary, in their contributing to the making of the estuary, oyster filter-feeding folds oyster body, mud bank, tidal flows, suspended sediment together into new meanings within the mangrove.

I argued that the nature of meaning changes depending upon the ontological model, and in this case, with these two variants of semiotic material ontologies. In the semiotic material context, this concerns how materiality makes itself meaningful. The layered model of biosemiotics, despite providing a material semiotic that does not require a human perspective, nonetheless endorses a hierarchical structure of nested scales that positions human symbolic capacities as evolutionarily more advanced than other semiotic capacities. Meaning here adheres to the substantive, sustained semiotic systems produced as layered semiotic material processes create stable meanings. It is associated with the stable outcome or emergent result. In the folded immanent version, meaning is in process within relations as they unfold and interfold. A reflexive turn is involved, but the folds are fractally dispersed, multiple, varied, and folding back into immanence. Where oyster flesh meets detritus-laden current, where mucus thread meets particle, along these sites of material semiotic interface, attentive material responding is occurring and differences are performed. Along the edge of becoming-different, as oyster flesh meets water current and the two affect each other into their becomings, meaning is being performed. Oyster semiotics supports a folded, but not a hierarchically layered materiality.

Through Part 2, I probed the material semiotics of the chronology of events occurring between a wallaby, the RRV, an *Aedes vigilax* mosquito, and myself, following the movement of viral code between and through three animal bodies. The viral trajectory defines a relationship among these bodies, meaningful for anyone who can read and follow the story. Mosquitoes connect the bodies of the human inhabitants of the surrounding suburbs to the mangroves of the Georges River. Every summer hordes of mosquitoes harvest blood for their babies from the mammals and birds in the areas near the river. That blood raises up the mosquito larvae that provide food for fish and invertebrates and the next generation of mosquitoes on the wing that feed the spiders and frogs. My body is deeply enfolded into the mangrove this way, but to describe the relation I have with the mangrove as, for example, "mediated by a mosquito", requires some storying work. I proposed a conceptualisation of relationality as storied within a semiotic materiality. Within a churning immanence, where intensities of interference or diffraction occur, discerning differences and connections such as those between a wallaby, mosquito, virus, and human, storying the continuities and discontinuities in such a way as to describe four beings connected through an invisible sickness, such a relational story is contingent upon many disparate factors. These include the histories and practices of epidemiology, diagnosis, and virology; the classification of insect genera and species, and the evolutionary history and particular daily circumstances of each of these entities. Our relation storied as the events enclosing a sickness is framed by that sickness. Relations are not just there; they must be narratively assembled.

Wallaby, mosquito, virus, and I orient towards each other in material semiotic obligations in numerous and diverse ways. Some of these ways involve human consciousness; others involve immune system responses, the evolution of senses, viral coding alignments, or feeding behaviours. Despite being attuned over many generations and within the more everyday temporalities of our encounters, our deep obligations towards each other are not readily interpretable as mutually beneficial, harmonious, or balanced. The example used in this research enabled my exploration to attend to relationality as ethically ambiguous. I argued that taking a semiotic material approach raises the importance of how relations are storied as an ethical matter of concern. Narrative strategies that might be useful for supporting relational narratives that encourage supportive co-obligations come into prominence, and several were considered here. These are strategies that experienced storytellers use to allow stories to remain unresolved, democratically shared, and carefully attentive.

In Part 3 of the book, my attention turned to difference and the walls of the Cooks River. My discussion was interested to understand the motivations towards investing such capacity and perpetrating the violences required to build and sustain these walls, and those of human exceptionalism. I identified an incentive in the desire to hold onto what had become regarded as a good, or a competitive advantage and considered the role of walls in enabling the possibility of winning as a narrative outcome. Rest, I proposed, is a key motivator in these activities. A desire to rest from relational work drives a search for sovereign security. Mangrove ways of walling are not like the colonial wall building practices along the Cooks River. Mangroves edge this river in ways that offer a more generous and accommodating alternative to the engineered embankments. They enact many of the same effects as walls, as they protect coastlines from erosion and provide sanctuary and storage, but they do these things in ways that generate lively, diverse, adaptable environments. I argued that their agency here in the urban sprawl, alongside the metal embankments, suggests a more-than-human resistance to the channels restricting and directing resource towards colonial narrative ambitions. I argued further that this mangal resistance can act as a regular reminder to humans in the catchment that these restricted flows are not the only possible ones. In performing as walls in porous and accommodating ways, I suggested that these mangroves offer clues about how to story relations thickly, how to practice difference in ways that avoid settling edges too firmly and in ways that incorporate indeterminacy.

Embankments are already being designed and constructed more in line with these mangal approaches, largely because experience has shown mangal walls are effective. However, while the eco-designed embankments of the Cooks River incorporate some of the porosities of mangroves, they are controlled by humans, designed to conform to human urban plans and conveniences, and control mangal possibilities. These eco-designed walls demonstrate some interest in inviting non-humans to make proposals, but little consideration of how to respectfully consider what they have to say. They indicate a promising beginning to multispecies design conversations in this catchment, while pointing to the possibilities and challenges of more genuinely living in more-than-human worlds.

Through exploring the tides, mosquitoes, and walls of Sydney's southern estuarine mangroves, I have ventured into the in-between and prised open some of its apparent strangeness, monstrousness, and impossibility. These have been repositioned as the ghosts of an anthropocentric horror story that is disturbed from its dreams of security by the co-vulnerabilities of endless interfaces between shifting encounters. I have argued that human exceptionalism and anthropocentrism have been justified by ontological accounts that find the meaningfulness of the world hierarchically layered. Instead, I have supported a turn to fractally folded material semiotic ontologies, which do not deny, but do not privilege human uniqueness. By embracing these ontologies, the key concern becomes considerations of how to interpret and respond into narrative coherences and their entangling obligations. This concern takes many semiotic material forms and is more-than-human and more-than-biological. By exploring examples of mangrove wall building and mosquito-human-virus relations, I have shown that engaging with this concern does not necessarily lead to harmonious, pleasant or mutually comprehensible outcomes. It can, however, support relations and becomings to remain sensitively attuned, to develop their own peculiar aesthetics, and to be sustained, adjusted, or transformed by the ongoing attentions of those oriented towards them. My discussion of the actions of a virus within a relational assemblage explained how semiotic material foldings incorporate elements that nudge unfolding narratives askew. The element of disturbance within regularity continues to demand attention to semiotic material narrative strategy. My argument is not for remaining always with immanence, uncertainty, and irresolution. Instead, my focus has been the tensions and challenges occurring between semiotic material narratives, as these inter-act, emphasising the in-between of attentive questioning, interpretation, and response as demanding and important. In discussing the metal embankments of the Cooks River, and the mangal walls that trouble them and offer alternatives, I have insisted that power matters within material semiotic narratives, that it seeks to hide, and that taking it into account is one of the continual activities of semiotic material narrative practice.

I hope this book contributes theoretically and methodologically to research seeking to decentre human perspectives and interests and supporting efforts towards living as more-than-human ecologies. I see this as important because I identify anthropocentrism as a central factor preventing human decision-making from considering other perspectives sufficiently to maintain global or local ecological complexity. This is an important and growing concern for many researchers within the environmental humanities and related areas. My focus upon mangal ecologies grounded this exploration in more-than-human perspectives, while my focus upon the in-between targeted an area of theoretical importance for making this perspectival transition. By combining these two elements, this book offers both theoretical tools and methodological examples that can support further non-anthropocentric, more-than-human approaches to research and practice.

Yesterday morning as I entered the mangroves at Lime Kiln Bay, the tide was high and the air was heavy. Perhaps because of the dark clouds and slight drizzle, I was the only human there. I stood for a while quietly on the pedestrian bridge, sharing some long view of the project with this place that has added so much to it. Raindrops played the surface, turning the reflections of the forest into a shimmery reverse of the forest. As my thoughts fell further into that dappled, flipped surface, a silver flash of fish came leaping through it, surprisingly high, returning smoothly into the water. Under me, an abundance of forest matter was leaving the mangroves, clumps of twigs, leaves, foam, and plastics slowly flowing towards the river and the sea.

Note

1 Douglas Hofstadter, *I Am a Strange Loop* (New York: Basic Books, 2007).

Reference

Douglas Hofstadter. *I Am a Strange Loop* (New York: Basic Books, 2007).

Index

Note: Endnotes are indicated by the page number followed by 'n' and the endnote number e.g., 20n1 refers to endnote 1 on page 20.